Digital Daily Management

Daily Management (DM) is a structured method that empowers managers and employees to establish a team-based, continuous improvement process. DM focuses on transparency by setting standards and key performance indicators, facilitating performance discussions centred on achievable goals, identifying deviations, and supporting on-site management of the plan-do-check-act cycle. Digital Daily Management takes these principles to the next level by leveraging digital solutions. This book simplifies the process, making it easy to understand and implement.

Digital Daily Management: Transforming Data to Improvement covers a pressing issue faced by production companies and provides a comprehensive roadmap for implementing efficient digital daily management. Drawing on extensive research and bolstered by a wealth of real-world examples, including four illuminating case studies from top industry players, this resource also offers valuable insights into the implications of artificial intelligence on this methodology. Readers will have access to downloadable templates that can streamline the implementation process. These resources include Excel templates for calculating return on investment, conducting a daily management audits and outlining specifications, as well as a detailed implementation guide.

Production managers, lean managers, operations managers, researchers, and students will appreciate the step-by-step instructions on how to introduce digital daily management to further strengthen an organization's problem-solving ability.

Digital Daily Management

Transforming Data to Improvement

Joachim Metternich, Marvin Müller,
Christian Hertle, Lukas Longard, and
Yuxi Wang

CRC Press
Taylor & Francis Group
Boca Raton London New York

CRC Press is an imprint of the
Taylor & Francis Group, an **informa** business

Designed cover image: SFM Systems GmbH and Sibylle Scheibner

First edition published 2026
by CRC Press
2385 NW Executive Center Drive, Suite 320, Boca Raton FL 33431

and by CRC Press
4 Park Square, Milton Park, Abingdon, Oxon, OX14 4RN

CRC Press is an imprint of Taylor & Francis Group, LLC

Originally published by Carl Hanser Verlag GmbH & Co. 3/24, in German, 171ppgs., 9783446477612

ISBN: 978-1-041-01017-3 (hbk)
ISBN: 978-1-041-01004-3 (pbk)
ISBN: 978-1-003-61268-1 (ebk)

DOI: 10.1201/9781003612681

Typeset in Palatino
by Newgen Publishing UK

Access the downloadable content: sfmsystems.de

Contents

Preface

Just like product innovations, the continuous improvement of processes and procedures contributes to the company's success. Process innovations are usually initiated and led top-down by management. This takes place as part of value stream planning or through individual improvement projects, for example, a line redesign. Individual, major changes are often the result. In contrast, there is the idea of a daily improvement in small steps, which is initiated and supported from the bottom up by employees – the so-called continuous improvement process (CIP), or Kaizen in Japanese.

This book addresses the digital support of the Kaizen framework of shop floor management – and not the digitalisation of the shop floor through Manufacturing Execution System (MES) systems or sensor retrofitting.

Kaizen originated in the quality movement of the 1950s and was coined by the American W. E. Deming in Japan. Unlike innovation, kaizen is about small, daily opportunities for improvement that are taken up and implemented at employee level. Toyota in particular exemplified the CIP philosophy very successfully, with the result that CIP was initially introduced in western automotive production in the 1990s under the term "shop floor management" (SFM). Since then, SFM has spread to many areas of work and business as a method of management at the point of value creation with the aim of fostering a culture of continuous process improvement, systematically using employees' skills for improvement processes and developing employees themselves.

The basic idea is for employees to systematically develop their problem-solving skills as part of problem-solving and the subsequent implementation of measures. SFM combines technical solutions (e.g., information and team boards) with organizational processes (e.g., the plan-do-check-act process) and the use of methods (e.g., systematic problem-solving). SFM can therefore be seen as a holistic management approach for improving processes and developing employees – and it is successful at that.

The disadvantages of SFM lie in the maintenance effort required for the boards and in the non-transparent progress of improvement activities, being tracked manually through paper based action plans. Data on the production status and ongoing measures must be recorded, distributed, and maintained manually. With the increasing digitalisation of many processes, more and more data is available automatically. In addition, interfaces and end devices create new possibilities for interaction between the production process and the teams, but also for organizing the CIP process itself. SFM offers an ideal environment for systematic CIP in production (but also in other areas of the company). Digital technologies can help to make the SFM less time-consuming, faster, and more effective. This is initially achieved by creating greater transparency by automatically providing data and key figures in real time for the dialogue conducted as part of the SFM. The increased transparency of the performance forms the basis for a performance management, in the context of which a team takes responsibility for its own process and develops ideas for its stabilisation and improvement. If a root cause analysis is necessary, digital tools can support structured problem-solving by providing quick access to data on defective products or systems. In addition, problem-solving methods can be supported by structured instructions and problems that have already been solved can be digitally documented so that they are available for further activities.

The goal of digital SFM is not to automate problem-solving by employees, but to improve their creative performance by providing data and known solutions.

This book shows how to select and implement a dSFM system to support the philosophy of CIP – and thus helps the company move forward. To this end, current challenges to the classic principles of lean production are first explained (Chapter 1). The basics of CIP (Chapter 2) and SFM (Chapter 3) are then discussed. Because only with the right understanding of methods can a dSFM be successful, this is also discussed in Chapter 4, which highlights the strengths and opportunities as well as the weaknesses and threats of the digitalisation of SFM. It also explains how different variants of dSFM affect a company. This builds up the knowledge needed to decide on a dSFM variant. Chapter 5 then guides the readers step by step through the necessary implementation steps and in chapter 6, four companies report on their experiences in the implementation process of different dSFM variants.

However, digitalisation can not only simplify existing processes in SFM, the resulting data basis also enables completely new evaluations. Chapter 7 concludes by showing which insights can be gained from the key figures and text data in SFM using artificial intelligence.

This book is the result of the research and practical projects of the Institute for Production Management, Technology and Machine Tools (PTW) at TU Darmstadt and its spin-off Shopfloor Management Systems GmbH (SFM Systems). The many years of experience in the design and introduction of classic and digital SFM reflect in this book through numerous practical examples.

We provide hands-on templates for your digitalisation project. You can download them from sfmsystems.de or scan the QR code below.

Amortization - Calculate your pay-back time (Excel)

Specifications and utility analysis for digital daily management software (Excel)

Implementation guide for digital daily management - drawio (PDF)

Daily Management Audit (Excel)

Acknowledgements

We thank our contributors for their field reports:

Julia Neubert, Project Leader of Operational Excellence Voith Group, Germany

Luca De Simoni, Product Owner at Pilatus Aircraft Ltd, Switzerland.

Stefan Munsch, Managing Director at Munsch Chemie-Pumpen GmbH, Germany

Volker Heuser, Former Head of Operations at Filtration Group, Germany

About the Authors

Joachim Metternich heads the Institute for Production Management, Technology and Machine Tools (PTW) at the Technical University of Darmstadt. After studying industrial engineering and completing his doctorate at the Technical University of Darmstadt. Prof. Dr.-Ing. Metternich spent 11 years working in industry for several German companies at national and international production sites. Most recently he was responsible for the global production system of a rail vehicle supplier before returning to the Technical University of Darmstadt. His research focuses on lean value streams and the use of digitalisation and machine learning to further develop lean production.

Marvin Müller is a customer success manager at SFM Systems GmbH, where he works on the development and implementation of digital daily management for industrial customers. Dr.-Ing. Müller completed his dissertation on the topic of text mining in digital daily management at the Institute of Production Management, Technology and Machine Tools (PTW) at the Technical University of Darmstadt in 2022.

Christian Hertle is managing director of SFM Systems GmbH and is responsible for Business Development. Dr.-Ing. Hertle studied at the Institute of Production Management, Technology and Machine Tools (PTW) at Technical University of Darmstadt for several years, specialising in daily management and its digitalisation for several years before he completed his dissertation on skills development in daily management in 2017.

Lukas Longard has been a senior engineer and doctoral student at the Institute for Production Management, Technology and Machine Tools (PTW) at the Technical University of Darmstadt. Dr.-Ing. Longard's research focuses on the development and implementation of data-driven approaches in digital daily management and performance management systems. As part of his work at the institute, he advises SMEs on the topics of digitalisation and lean production.

Yuxi Wang (MSc) is a research assistant and doctoral student at the Institute of Production Management, Technology and Machine Tools (PTW) at the Technical University of Darmstadt. Her main research area is data-based decision support systems for deviation management on the shop floor.

1

Lean Production Today

ABSTRACT Solving problems sustainably to avoid waste remains the success factor of lean production. Changing challenges for manufacturing companies as well as new opportunities, for example, through digitalisation, mean that established lean methods also need to evolve without abandoning the fundamental lean philosophy.

Lean production at the Japanese car manufacturer Toyota arose from the emergency situation that prevailed in post-war Japan. A shortage of resources and restrained demand led to the need to avoid waste and only produce what was necessary. In production, methods such as just-in-time, pull, takt, and flow were developed and refined [1]. Over the years, lean production has been continuously expanded into a management system. Today, we understand lean production as a strategic approach to operational excellence in the dimensions of time, quality, productivity, and flexibility. It is based on values and consistently exemplified behaviours such as local leadership, respect, teamwork, and the development of excellent employees (Figure 1.1).

These values and behaviours, including the demand for continuous improvement, can be applied to almost all company areas.

The potential of lean production was recognised worldwide in the 1990s, and it initially found its way into the automotive industry. Other industries quickly followed, and as a result, many companies have achieved significant efficiency gains and competitive advantages over the decades in various lean waves.

Since the 2008 financial crisis, the impression has arisen that the market environment, technological possibilities, and customer requirements are changing ever more rapidly and, in some cases, radically. Added to this are the changes in the labour market regarding labour supply and employees' expectations of meaningful employment. In the following, we explain some of these changes and what they mean for the further development of lean production [3].

DOI: 10.1201/9781003612681-1

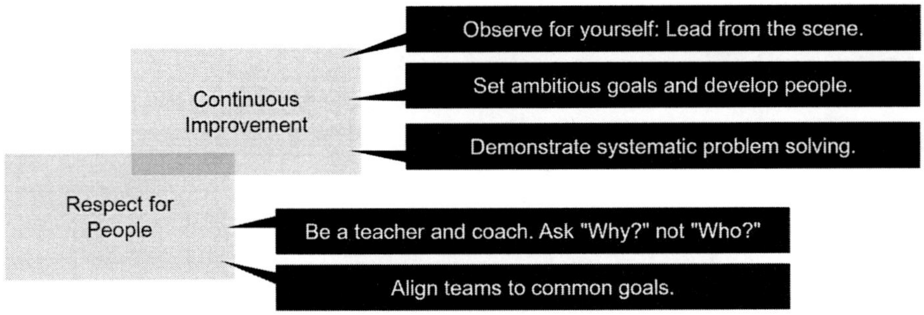

FIGURE 1.1
Lean values and behaviours using Toyota as an example [2].

1.1 Individualisation

The customised solutions sector is growing in many markets. Companies that successfully provide this customisation in the shortest delivery time can gain a visible competitive advantage. To resolve the conflicting goals of customisation and response time, the classic waste-based approach and, thus also, the classic value stream design falls short. Instead, the scope of possible product specifications must match the capabilities of the company's own order processing. The ordering process must be designed to protect the following processes (customisation development, procurement, work preparation, and production) from unwanted combinations of product features. Quotation processes, customisation development and work preparation, should be automated using digital solutions wherever possible. The flow of information must be designed consistently and effectively so that the voice of the customer can be heard at every workstation. Worker assistance systems can help to implement standardised work at individual workstations despite individual work steps – just as in mass production. Standard work not only leads to an increase in efficiency but also forms the basis for deviation management, which is very important in daily management (DM) (see Section 2.2 in Chapter 2).

1.2 Disrupted Supply Chains

In the past, international sourcing or the relocation of production to low-wage countries was largely driven by cost reasons and, at times, by local regulations. Increasing sales were achieved by tapping into new, international sales markets. The fragility of supply chains has been made painfully clear by the events of recent years. The closure of harbours, international sanctions, accidents on important trade routes, or direct material shortages has led to the disruption or complete collapse of supply chains that have been successful for a long time. In the short term, inventories, which initially represent waste from a lean perspective, take on a completely different meaning – they become "protective buffers". To a greater extent than before, lean production must incorporate risk into the design of such strategic inventories. Toyota also demonstrated this itself in 2021. The Japanese car manufacturer was the market leader, with the most cars sold in the USA for the first time. Strategic stocks of chips were built up early, and unlike the local car manufacturer General Motors (GM), Toyota was still able to deliver [4].

The excessive focus on costs in sourcing has led to a dead end. In the medium and long terms, companies must increase their resilience and self-sufficiency to deliver and protect entire economies from blackmail. Increased cooperation with countries that share similar values will be the result. These activities towards more localisation in the semiconductor, battery and solar energy sectors are already clearly visible.

1.3 Climate Neutrality and Sustainability

The pressure on companies to reduce CO_2 emissions and manufacture resource-saving products comes from legislators, direct customers, and end consumers. In just a few years, the ability to show a CO_2 footprint for products will determine whether or not a company is allowed to deliver. End-to-end traceability solutions must be established to ensure the necessary transparency. The subsequent regulation and reduction of specific CO_2 consumption can take place via key figures and targets in deviation management. These CO_2-related deviations will be part of DM in the future. The same applies to the conservation of other resources, such as materials and operating resources. The concept of the circular economy offers a comprehensive set of tools for this that goes far beyond recycling. In order to "save" the added value of already manufactured parts and return them to the production cycle, product structures must first be changed. The fact that the product structure also determines the production structure leads to new processes in the retrieval, disassembly, cleaning, analysing, and reconditioning of products and their components. This will also require new planning and management methods. In order for these processes to be more economical than the production of new parts, lean production must offer new solutions – which can be developed by DM.

New business models also need to be developed to bring parts that have already been used back onto the market along with new parts as part of new products. As such, the understanding of "new" and "used" will change.

1.4 Demographic Change

The imminent retirement of the baby boomers from the labour force will foreseeably lead to a decline in the working population. This will be a leading cause of the future shortage of skilled labour. The situation is exacerbated because the number of new entrants to relevant engineering degree programmes has been falling for several years. While the need for replacements in STEM-related jobs is increasing, the future supply is decreasing. In addition, an increasing proportion of people are working part-time, with later entry into the labour market and the individual rebalancing of private and working life. It is clear that massive increases in productivity are required if economic output and, thus, prosperity are to be secured. Unlike in the past, this also requires the automation of activities that can be categorised as mid-level skills, such as planning and control activities.

It is also clear that the skills required of employees are changing faster than before and may change several times in a working life. Continuous training of the workforce will become an important competitive factor. This will be more work-related or even work-integrated than in the past, giving the development approach of the continuous improvement process (CIP) even greater significance. For example, the DM can address this by assigning employees specific tasks in which they can gradually develop their existing level of expertise.

1.5 Digitalisation of Processes and Products

The wave of digitalisation under the buzzword "Industry 4.0" has raised many questions, particularly with regard to lean production. Industry 4.0 aims to create processes that optimise themselves based on data and its analysis. Lean, on the other hand, focuses on people and their problem-solving skills. We now have powerful tools available for CIP based on data and artificial intelligence (AI). The low-cost retrofitting of sensors, standardised interfaces, and the free configurability of dashboards make it easier to create transparency and identify deviations. This is very much in the spirit of lean. Without in-depth expert knowledge, freely available libraries of powerful AI algorithms can be configured and integrated into operational monitoring processes. This enables predictive approaches in quality assurance and maintenance to identify deviations directly based on process data before performance losses occur or defects develop.

While Industry 4.0 emphasises flexibility by breaking down fixed process links and fixed sequences, lean is based on standards as the foundation of CIP.

Standards are characterised by the fact that there is a time specification per work step ("takt"), the order of the work steps is fixed ("sequence"), and the number of parts must not exceed an upper limit ("inventory"). In many practical examples, companies show that combining both ways of thinking is possible. With products that independently search for the next free work system, that are information carriers and, as such, adapt the work system (e.g., through automatic programme generation or parameter selection), and that automatically initiate the configuration of work instructions and ensure material withdrawal through appropriate pick-by technologies, the lean idea of standard work can also be implemented in small series or single-item environments.

In addition, data analytics methods enable utilising data from Enterprise Resource Planning (ERP) systems, Manufacturing Execution System (MES), and other reporting points to develop further established lean methods. One example of this is the value stream method. With the help of data we can

- find product families based on product–process matrices using cluster analysis,
- discover unknown process links through "Process Discovery",
- determine cycle, setup, and idle times as well as inventory changes by analysing reporting times,
- define and plan the scope and capacity of work systems using mathematical models.

However, digitalisation not only impacts internal company processes, the market is also changing rapidly and sometimes disruptively. Additionally, not only have sales and marketing channels shifted to the internet, but digital business models can also increase the value created for the customer. Traditional business models were focused on individual purchase transactions and provided hardly any added value in the usage phase. Whereas in the past, the purchase transaction with the transfer of a good and the associated payment took centre-stage, digital business models also focus on the usage phase by providing the user of a product with information on improved utilisation by analysing usage data. This can involve process settings, quality predictions, improvements, or maintenance recommendations. In addition, the connectivity of machines and their components enables new payment models that may replace traditional leasing. Fixed instalments can be replaced by payments based on the component's wear and tear. However, many companies are still finding it difficult to explain the benefits of digital services to their customers and to price them accordingly.

Regardless of whether we are talking about the digitalisation of processes, products, or services, the demands on employees' digital skills continue to rise. More than ever, they need to be able to implement digital solutions, collect data and information, and ultimately interpret them so that the quality of decisions increases.

dDM is the ideal daily routine to develop these competencies and, at the same time, drive forward the CIP.

References

[1] Liker, J. K.: *The Toyota Way. 14 management principles of the world's most successful automotive group*. Munich: FinanzBuch-Verl. 2006.

[2] Liker, J. K.; Meier, D.: *The Toyota way fieldbook. A practical guide for implementing Toyota's 4Ps*. New York: McGraw-Hill 2006.

[3] Abele, E.; Reinhart, G.: *Zukunft der Produktion. Challenges, research fields, opportunities*. Munich: Hanser 2011.

[4] Reiche, L.: Toyota knocks GM off its throne in the USA. Internet: www.manager-magazin.de/unternehmen/autoindustrie/toyota-stoesst-general-motors-in-usa-vom-thron-a-5469314f-6628-410e-9d29-a649fe265a7a. Accessed on 21.01.2023

2

The Continuous Improvement Process in Daily Management

ABSTRACT Stable and standardised processes are the basis for systematic improvement. A standard performance can be defined for them, and process specifications can be made. When compared with the actual process performance, these process specifications allow deviations to be recognised and trigger the improvement processes in DM. In the following, the terms "Kaizen", "CIP", and "process specification" are explained in the context of DM.

2.1 Classification of Kaizen, Continuous Improvement, and Daily Management

This section explains how improvement projects differ from the CIP approach in DM.

Kaizen means literally "change for the better". The term is used to describe activities of very different scopes (Figure 2.1).

Improvement activities at the value stream level (flow kaizen) address the waste that arises from the interaction between several departments, machines, and workstations, which, for example, results in a lot of transport, long throughput times, and high inventories. As part of value stream management, major increases in efficiency can be achieved by creating flow islands, setting up pull connections and synchronising material and information flows. Waste that affects individual workstations or machines and systems can be tackled through improvement projects (Point Kaizen). Examples include the quick changeover of a machine tool, the reorganisation of a warehouse to reduce walking distances or the ergonomic redesign of workstations. With this approach, there are usually clear targets (e.g., reduce changeover time from X hours to Y minutes). Based on a waste analysis, a project team develops a target state to be achieved through a sequence of improvement activities. These are summarised in a project plan with clearly defined responsibilities and timescales and are continuously monitored. If a

DOI: 10.1201/9781003612681-2

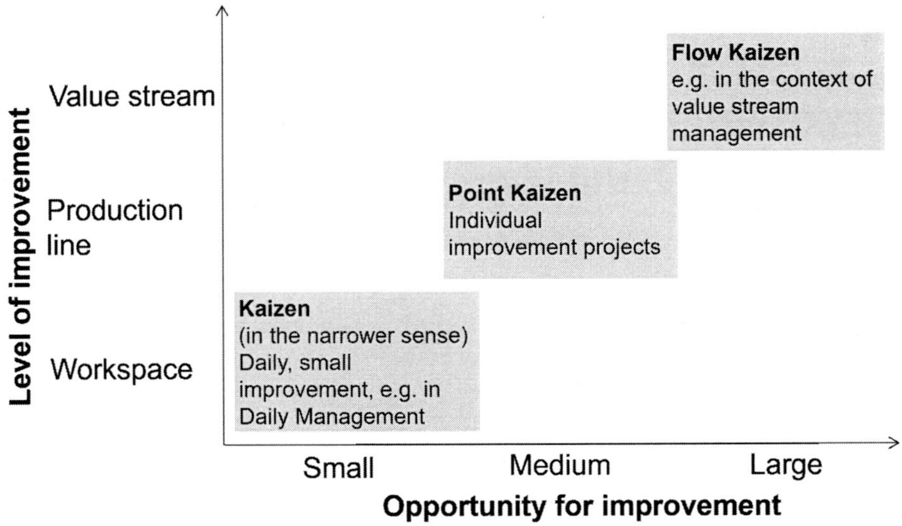

FIGURE 2.1
Classification of DM in the context of Kaizen. (Generated by authors.)

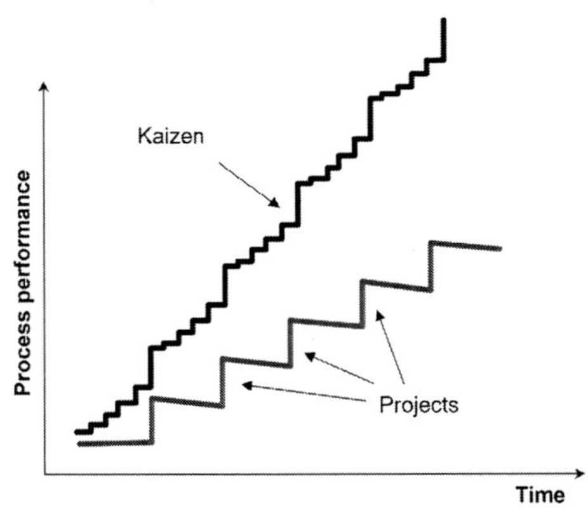

FIGURE 2.2
Effects of projects and CIP on process performance. (Generated by authors.)

new process is defined or a new technology is introduced, it is necessary to ensure that the status achieved is continuously monitored. If this is not done, there is a risk that some of the improvements achieved will be lost again (Figure 2.2).

In a narrower and more common understanding, Kaizen is understood as a continuous improvement process. As part of the CIP, the workforce continuously identifies opportunities for improvements and problems and solves them on-site using simple means. The individual steps are small, inexpensive, and unspectacular but compound.

	CIP	Project
Effect	Small, unspectacular steps	Major, radical changes
Scope	Continuous	Unique
Participation	All as a collective	Individuals as experts
Approach	Stabilise, standardise, and then improve	Destroy and rebuild
Trigger	Existing knowledge	Innovations
Expenditure	Small investment	High investment
Focus	Improving people and processes	Technology as a driver

The CIP is ideally suited to tapping into employees' knowledge of their work processes and systems for stabilisation and gradual improvement. However, the classic CIP also requires a suitable environment to ensure its initiation, continuity, and targeted execution. In addition to cultural aspects such as trust and fault tolerance, having an appropriate infrastructure and organisation also plays a role. DM provides an excellent framework for initiating the CIP repeatedly, allowing it to run in a targeted manner. It provides the methodological, temporal, and spatial context to bring together the right stakeholders and knowledge bearers, enabling them to identify and prioritise new opportunities for improvement daily, translate them into actions, and track their implementation.

2.2 Prerequisite for Continuous Improvement: Process Specifications

The prerequisite for continuous and targeted improvement is that employees can recognise at a glance whether their process is in a normal state or has a deviation. The basis for this are realistic process specifications. Deviations from these specifications also give managers starting points for a qualified performance dialogue. This section shows the necessity and variants of process specifications, including the possibilities of using them in CIP.

2.2.1 Necessity of Process Specifications

Without process specifications, it is not possible to objectively recognise whether a process is in a normal state or whether there is a deviation (Figure 2.3).

In a highly fluctuating environment without targets, deviations are not systematically recognisable. There is a "better" or "worse" – but there is no comparison level for "too bad" or "good enough".

> "Without standards, there are no problems, only opinions."
>
> Nate Furuta, President and CEO of TOYOTA Boshoku [1]

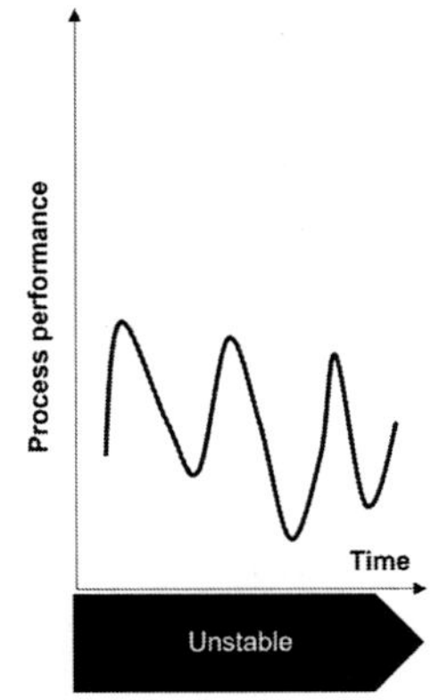

FIGURE 2.3
Without the specification of an expected process performance, no deviation is recognizable. (Generated by authors.)

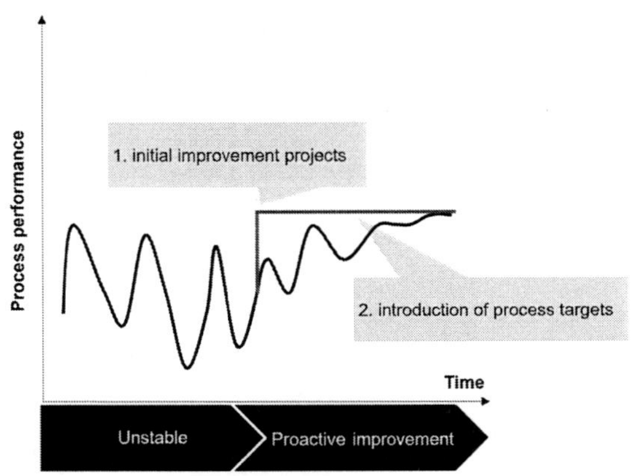

FIGURE 2.4
Change in process performance over time when using improvement projects and DM. (Generated by authors.)

In order to set realistic targets, the range of fluctuation for a process must first be proactively reduced. These target values reflect the normal performance of a process. They, therefore, also provide the comparison level for the current process status. Hence, it is possible to objectively determine whether a process is running "well" or "poorly" and whether improvement activities are necessary. It is no longer just opinions that count but the measurable achievement of the process specifications (Figure 2.4).

2.2.2 Process Specifications: Standards and Key Figures

Process specifications form the trigger for the CIP within the framework of DM. It is, therefore, worth taking a closer look at the various process specifications.

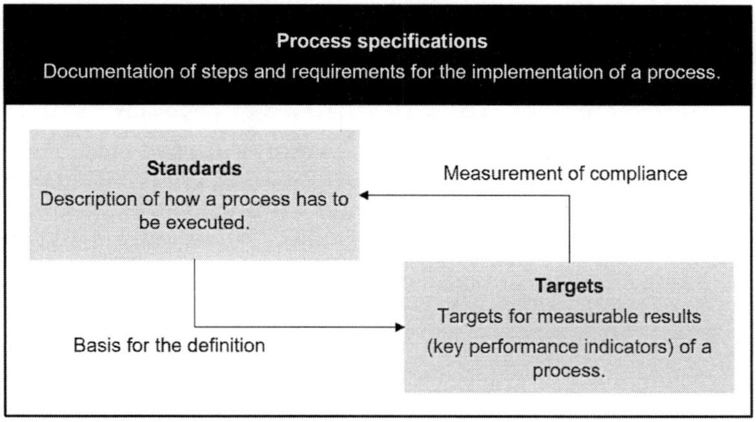

FIGURE 2.5
Relationship between process specifications, standards, and targets of key figures. (Generated by authors.)

In this book, we divide the process specifications into standards on the one hand and the resulting targets for specific process Key Performance Indicator (KPIs) on the other hand (Figure 2.5).

Targets can be derived from a standard process, e.g., compliance with the specifications results in a process time of 1:30 minutes. This can then be issued and measured as a target for the process time. If it is exceeded, it indicates the standards have not been met.

In terms of standards, we distinguish between standard work, technical process standards, and general labour standards. Standard work defines the work sequence, times, and stocks at workplaces defines the takt time, the work sequence and the standard inventory. The associated specifications can be found in the following documents:

- The classic work instruction defines how and with which aids activities are to be carried out at a workplace and provides information, for example, on which aspects of the activity are particularly critical and deserve attention.
- The Standardised Work Chart defines the time execution time for a workflow (takt), the correct order of worksteps (sequence) and the amount of material at the workplace (inventory).
- The Work Element Combination Chart shows the distribution of work between man and machine, e.g., in a multi-machine operation.

In addition, technical process standards must be observed, which can be found on machines and systems or in programmes or settings. Examples of technical standards are pressures, temperatures, quantities of consumables, oil types, deadlines, etc.

And last but not least, general work standards apply to a work area or factory, e.g., the procedure for shift changes or material supply, statistical process control, 5S audits, etc.

Key figures and their targets should be defined to meet important operational standards. For example, target quantities can be calculated based on target times and overall equipment effectiveness (OEE) and used within a shift for a target/actual comparison (see Section 3.2). Falling short of the target at the end of a shift always leads to a deviation and triggers a CIP in the DM. Key figures that are not derived from a process specification, e.g., cost reduction of 5% across the board, may not be achievable and are therefore not suitable for the DM.

It is, therefore, necessary to create process specifications carefully so that they fulfil their function of controlling the CIP.

2.2.3 Create Standards

Standards are the basis for continuous improvement. But how do you arrive at a standard? The basic process is shown in Figure 2.6.

The first step is to fundamentally understand the process to be standardised and to identify waste and variability. A range of lean methods are available for this purpose. The causes must be understood together with those involved and eliminated through improvement activities. The next step is to describe the flow of the new process (e.g., using standard worksheets, work instructions, etc.). Employees are often not trained intensively in the new processes, leaving questions unanswered and the standard unrefined. Finally, it is also a matter of anchoring and following up on new behaviours.

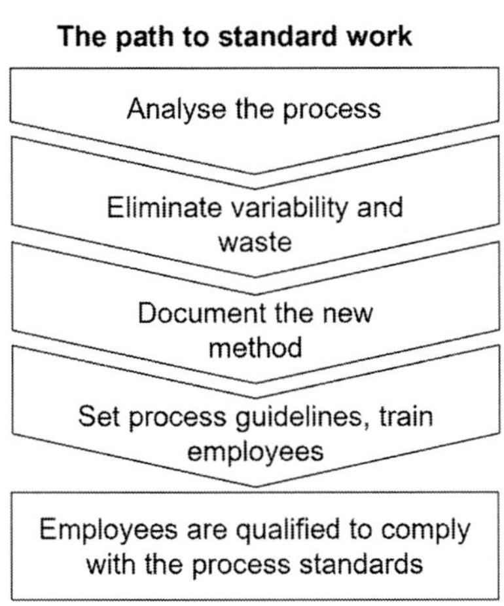

FIGURE 2.6
The basic process on the way to standard work. (Generated by authors.)

Sometimes, this requires the long-term presence and support of team leaders and managers on site. The result is a production environment in which employees deliver a consistently high number of units of the required quality on stable machines and systems with consistently available material, capable production processes, and clearly defined work sequences [2].

In many cases, it is still too early to introduce standardised work due to unstable processes. Unstable processes do not always run poorly; under certain circumstances, they can also deliver high process performance. However, excessive fluctuations in key figures (quality, output, availability, etc.) mean that process specifications cannot be consistently met. This would overtax a systematic CIP such as the one we want to pursue in DM. It is important to recognise this variability and eliminate its causes in advance, e.g.:

- Equipment failures
- Widely varying process times
- Lack of technical stability (e.g., pressure and temperature)
- Unsuitable process (e.g., damage due to transport)
- Missing material
- Quality defects
- Uneven scheduling of orders

> At a packaging machine manufacturer, orders from the following month are brought forward if the monthly planned turnover has not yet been reached. The consequence is that chaotic conditions regularly arise at the end of one month and a standstill at the beginning of the next. These homemade fluctuations, contrary to the CIP, could easily be avoided.

To achieve this, various strategies and tools are used in both the lean and digitalisation areas (Figure 2.7) [1].

A detailed explanation of these strategies and methods would go too far here. It should be noted that the standardisation of wasteful and unstable processes does not make sense and cannot form the basis for deviation management.

2.2.4 From Standard to Process Target

Process targets can be derived from the previously defined standard work conditions in terms of takt, sequence and inventory: for example, daily targets for the number of units to be produced can be derived from the defined customer takt. Here is an example from the assembly of an air shut-off valve.

Strategies

- Eliminate "major" sources of waste
- Improve operational availability
- Isolate, consolidate and reduce variability (e.g. no material provision by workers, instead consolidate and reorganise replenishment activities and assign it to a line feeder associate)

Tools within lean production

- Recognising waste, e.g. "Stand in the circle"
- Workplace organisation through 5S
- Quick changeover (SMED)
- Preventive maintenance (TPM)
- Smoothing of order scheduling

Digitisation approaches

- Process Mining
- Predictive maintenance based on condition monitoring

FIGURE 2.7
Strategies and tools for creating basic stability. (Generated by authors.)

The assembly workstation is first fundamentally improved with the employees on site in terms of ergonomics and equipment so that the process can be carried out in the specified time reliably. The corresponding documents for the standard work are then created (procedure according to Figure 2.6): The work instruction with the description of the activities and the workflow diagram with the corresponding times of all work steps (Figure 2.8).

The left-hand column lists all the work steps required to assemble a module, and the right-hand part of the document assigns a time to each work step and visualises it using bars. The visualisation shows that the sum of all individual times results in a total time of 2 minutes and 10 seconds per assembled module.

This cycle time of 2:10 min can now be used to determine a target number of pieces per shift, which serves as the target value for deviation management. To do this, the working hours per shift is divided by the cycle time:

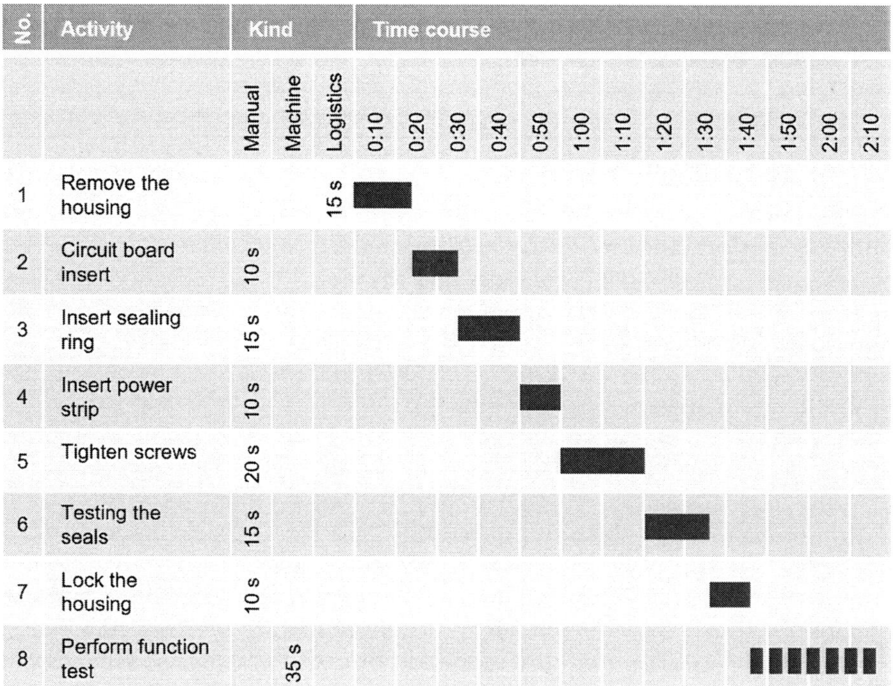

FIGURE 2.8
Workflow diagram of the installation of an air shut-off valve (simplified). (Generated by authors.)

Shift duration in minutes: 480 min
Break times: 30 min
Target number of pieces (theor.): (480 – 30) min / 2:10 min = 207 pieces.

However, as the line still has a certain degree of instability, this target would initially be almost impossible to achieve, and every shift would generate a deviation. We therefore recommend including the (in this case known) current system availability of a line in order to adjust the target number of pieces according to the current performance (in our example, the system availability = 70%). This results in the target number of pieces:

Target number of pieces (practical) = 207 pieces × 0.7 = 145 pieces

Deviation management should start with a shift target of 145 to determine whether it is sensitive enough to generate deviations and not too demanding so that the number of deviations does not overwhelm the subsequent CIP. The more stable the process develops over time due to the ongoing CIP, the fewer deviations can be expected. The team must now gradually increase the target number of items so that deviations continue to trigger the CIP.

For example, targets for DM can be defined on the shop floor – but key figures and target systems must also be harmonised with the company's objectives.

2.2.5 Key Figures and Targets

Key figures and associated targets are important elements in DM. Deviations from key performance indicators trigger analyses and measures – in other words, they control what employees and managers focus on. Therefore, key figures, targets, and how deviations from targets are handled must be carefully selected and designed.

Key figures must align with the company's objectives at all levels and not contradict each other across departments.

> There is a classic conflict of objectives between machine utilisation (measured by the key figure OEE) and throughput time. Large batch sizes reduce downtimes due to setups and improve the OEE. However, large batches also lead to longer waiting times until another product can be machined. This can lead to waiting times in assembly and ultimately increases a company's response times to customer orders and additionally inflates inventory levels and tied capital.

Employees and managers utilise their knowledge and creativity in line with formulated objectives, especially if financial incentives are linked to achieving objectives. If the objectives in different areas of order processing do not complement but rather contradict each other, many efforts get wasted and the company does not progress overall (left, Figure 2.9).

If managers are measured against uncoordinated targets in their area, it encourages self-optimisation in their own "principality". Shifting problems to other cost centres is a successful strategy – but one that does not contribute to improving the company and customer value. In contrast, company-wide harmonised process goals focus the problem-solving ability of employees on overarching, customer-relevant processes (right, Figure 2.9).

The challenge in developing objectives is to concretise the abstract corporate objectives into SMART objectives for the team level. Figure 2.10 shows the importance of "SMART" goals.

These rules for key figures are often known and appear simple but are repeatedly violated in practice. The following examples illustrate the negative effects on DM when these rules are disregarded.

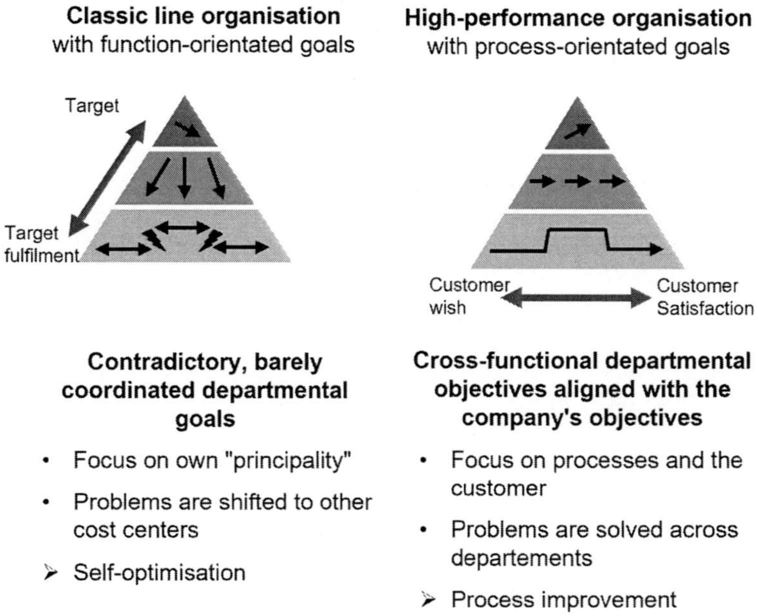

FIGURE 2.9
Change in process performance when departmental objectives are harmonized. (Generated by authors.)

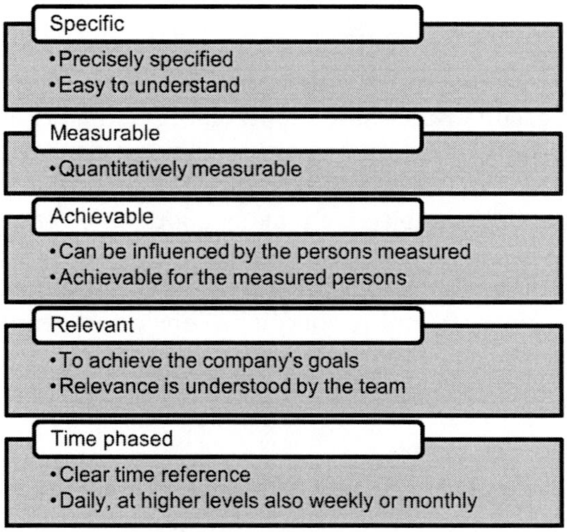

FIGURE 2.10
Definition of "SMART" key figures. (Generated by authors.)

Specific

The exact definition, data basis, and calculation methods of key figures are often unclear or unknown to all users. As a result, identification with a key figure and the associated target suffers. In addition, different areas that report the same KPI are not comparable with each other.

> At an automotive supplier specialising in sheet metal forming, all orders were evaluated using a performance index (PPI). If the PPI was below 70%, the machine operators involved had to report a very large DM round. This meant that under-fulfilment was directly linked to a perceived penalty. However, the key figure itself was problematic, as hardly anyone fully understood it, as various weighting factors were used to measure whether "rough setup", "fine setup", process speed, downtimes, and quality met the order-specific targets. The machine operators did not realise how much they could influence this figure through their actions. For the process engineers, too, many factors were unsuitable for searching for causes and defining measures. This was only recognised as a problem years later, and the DM was rebuilt. The machine operators now have clear output targets and address deviations in their own DM meetings. The PPI is no longer used

Measurable

For clear and, above all, objective deviation detection, it is necessary to use quantifiable and easily visualisable key figures. Qualitative assessments such as customer satisfaction (e.g., expressed in subjective ratings or comments), employee motivation (e.g., expressed in qualitative feedback), or team cohesion (e.g., expressed in a subjective scale of cooperation) are not suitable for daily deviation detection. As these are subjective assessments, there is a lack of comparability (across teams and over time), and it is difficult to derive specific measures. Such "soft" topics should be considered regularly and shared within the team. In daily DM – objective, quantitative, and easily understandable key figures are required.

> A Japanese start-up that offers dDM systems for Toyota's supplier base uses very simple figures: Cycle time, downtime, and number of parts. The machine operators can understand and influence these values directly. They are well suited for deriving and testing improvement measures. The data is collected by magnetic switches, light sensors, and light barriers directly on the machines, e.g., on the machine door or on the Andon (machine status) lamp.

Reachable

Repeating failure to achieve the target leads to frustration in the team and, in the long term, to the team no longer trying to achieve the target.

One example of a lack of reachability is the presentation of unit costs on the shop floor. Material prices, hourly machine rates, and even overhead rates are included there – all values that the team cannot influence. The team can only work on its own working time per unit or order and should, therefore, only be measured against this.

A working group at a manufacturer of measuring instruments was given more autonomy to fine-tune its orders and thus assume responsibility for meeting delivery reliability. The group vehemently objected to this because, according to the group, an essential semi-finished product had not been available in sufficient quantities for years, and the key performance indicator was, therefore, not achievable. The production manager responded to this justified criticism by stating that the problem had already been solved years ago. The situation was resolved by determining and tracking the frequency with which the affected semi-finished product was missing as a KPI. Measures to improve delivery reliability could then be derived from the development of this key figure.

Relevant

While unachievable goals lead to frustration, irrelevant goals lead to waste. For example, time is wasted collecting and discussing an irrelevant figure, and problem-solving capacity is wasted on irrelevant activities. Non-essential key figures in the team are often the result of standardised reports to management or historical definitions that have long since lost their relevance to the current process.

During a key performance indicator workshop at a manufacturer of pneumatic cylinders, it was determined that a general target of 70% for the KPI Overall Equipment Effectiveness (OEE) for all machines was not reasonable. While a saw cutting parts would generate overproduction if it met this target, a milling machine for cylinder bases would not be able to meet demand. It was therefore decided to pursue OEE primarily for bottleneck machines and to set individual targets for each of these.

Time-Related

A short cycle for data collection is crucial for the success of problem-solving. Deviations in key figures relating to the current or last shift can easily be documented, tracked, and analysed. Too many individual processes and external influences are summarised in weekly or monthly values in order to identify problems efficiently. In extreme cases, deviations are only recognised weeks after they occur. It is then no longer possible to analyse the causes. Evaluations based on weeks or months can nevertheless be used to prioritise recurring deviations and associated countermeasures.

> Monthly scrap cost analyses are interesting for controlling the quality department. However, a shop floor team cannot work using this figure on a day-to-day basis.

To summarise, key figures for effective daily management

- Must be linked to the overarching corporate objectives,
- Should measure aspects that are decisive for or prevent the achievement of objectives,
- Should be easy to understand and easy to measure,
- Should be updated at least daily to capture abnormal conditions and underlying causes as long as they are fresh.

A target must also be defined for each KPI in order to determine whether a process is in a normal or abnormal state.

The targets should

- Reflect the expected process performance under normal conditions,
- Carefully be defined to uncover enough problems to lead to improvements,
- Not be too demanding so that every day becomes a "red" day.

If the objectives for a production line or a team are set consistently as part of an target deployment process, then achieving the objectives of a value-adding process leads directly to the company's objectives being achieved. This enables teams and employees to make a measurable contribution to achieving the objectives of the entire organisation through their work.

The challenge in setting targets is to achieve constant, systematic, and sustainable improvements over time – continuous improvement is a marathon, not a sprint!

2.2.6 Examples of Key Figures at Different Hierarchy Levels

The challenge of target development lies in translating strategic targets at the company or business unit level into key figures and target values that enable coordinated and target-orientated action – from the business unit to the plant and department, down to the production line and the individual teams.

Strategic target categories can be

- Turnover and profit,
- Productivity increase (costs),
- Reduction of throughput time (time),
- Quality leadership (quality),
- Inspire customers (innovation).

The first step is to define specific values for target categories (Figure 2.11).

Furthermore, this should not be a blanket "target dictate", a top-down approach. Rather, it is about recognising which goals are sensible and achievable. Formulated correctly, a goal first triggers the desired process improvement before the target value is achieved through this improvement. It is, therefore, not just about achieving a goal but, above all, about the desirable improvements through which it can be achieved. This coordination of desired results and underlying process improvements is part of the target dialogue between the hierarchy levels.

The achievability and the way how to achieve a target must, therefore, be taken into account when breaking it down into work areas and teams. As a rule, key figures become simpler the closer you get to the workshop level.

While key figures, such as OEE, productivity, and error rate, are still available as ratios up to the department and group management level, at line and team level simple key figures should be used that can be updated as promptly as possible to enable a continuous comparison of targets and actual values – such as quantities, setup times, or the number of rejected parts. These metrics are usually absolute figures due to their immediate traceability and easy comprehensibility. They should be selected and provided with targets so that the fulfilment of a shift's target ensures that the daily quantity of the line is

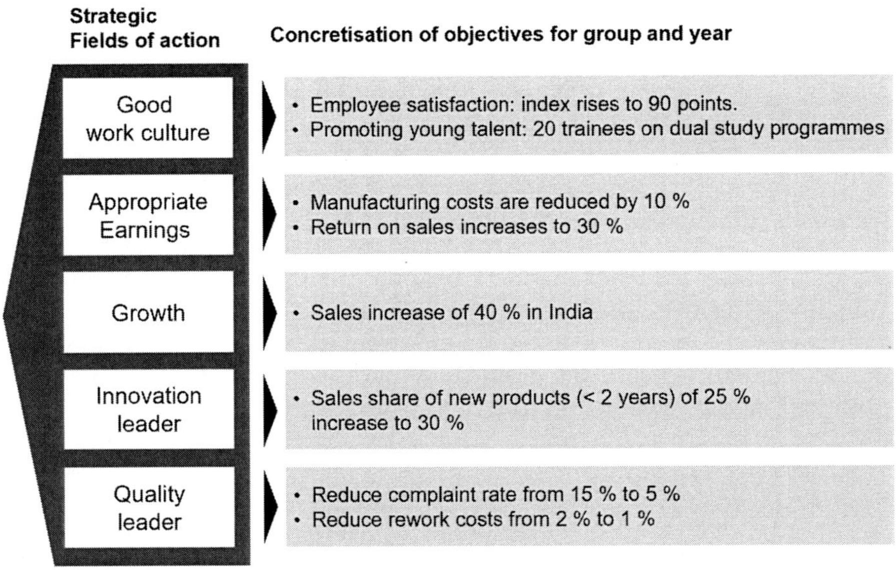

FIGURE 2.11
Concretisation of annual targets for a group of companies by the management. (Generated by authors.)

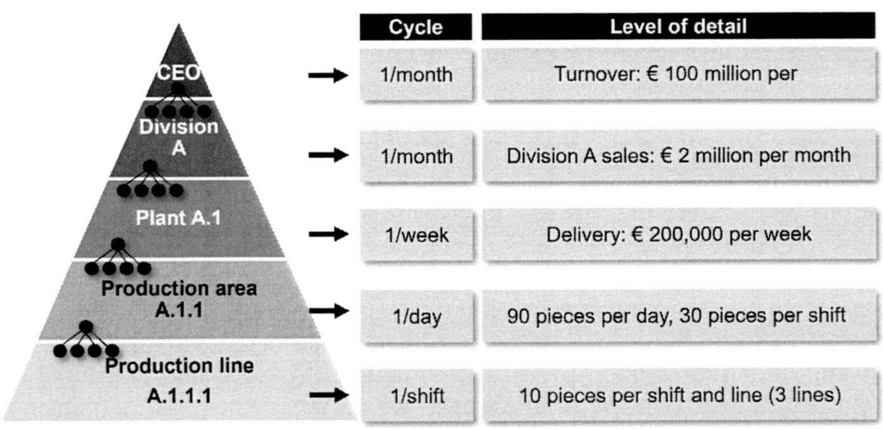

FIGURE 2.12
Translating a company's sales target into unit sales per shift and customer cycle per line. (Generated by authors.)

met and that a department ultimately fulfils the set productivity target. The example in Figure 2.12 shows how the monthly sales figure for a company is ultimately used to determine the shift performance of a line and, if applicable, the customer cycle for this line.

Appropriate target development ensures that the employees on a shift can contribute directly and measurably to fulfilling the company's targets. In other words, if the lines in each shift fulfil the specified target for a (suitably selected) key figure, the plant and, ultimately, the entire company will achieve its target in the associated field of action.

If generalised targets are not translated into realistic, area-specific goals, employees and middle management are unlikely to accept them and, as a result, will not prioritise working on them.

> It turned out that target values had been set across the board for all production plants and areas, regardless of the potential for improvement that still existed. As a result, the plant was given a blanket productivity target of 5% for this financial year, even though a significant leap in productivity had already been achieved in the previous year. This target was therefore not accepted by either local management or workshop management. It turned out that target values had been set across the board for all production plants and areas, regardless of the potential for improvement that still existed. As a result, the plant was given a blanket productivity target of 5% for this financial year, even though a significant leap in productivity had already been achieved in the previous year. This target was therefore not accepted by either local management or workshop management.
>
> The target deployment process has failed here – the various hierarchical levels should find realistic target values based on the company targets and local processes.

In the example above, pieces per shift and per day were used. While such an example and the concept of "standard work" may initially suggest large-scale production, a CIP based on process specifications also applies in a multi-variant environment, e.g. if the reported productive time for a product mix is compared with the total target time.

2.2.7 Process Specifications in a Multi-Variant Environment

In order to select and detail process specifications appropriately, it is important to first be clear about your own order entry point. In small series and customised production, it is necessary to control the earlier entry point of orders and, therefore, frequently change products and specifications (Figure 2.13) in the selection and definition of the specifications.

The effort required to describe individual work steps is no longer economically justifiable with a greater number of variants, as the frequency of repetition is lacking. Nevertheless, even beyond large-scale production (or make-to-stock), there are approaches to utilising standard work in a meaningful way with reasonable effort.

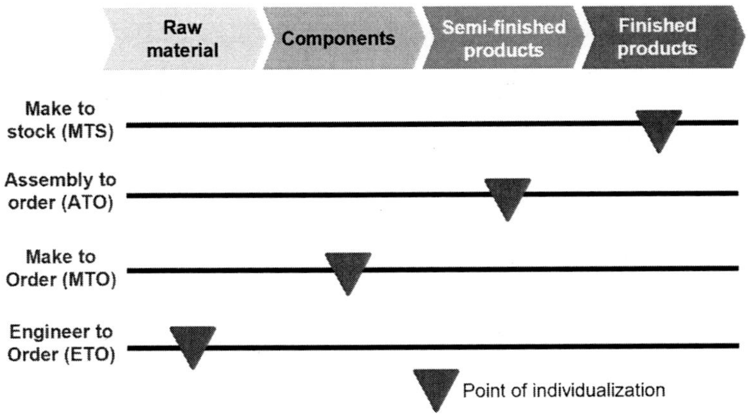

Factor	ETO	MTO	ATO	MTS
Delivery times	Long	Medium	Short	Very short
Lot size	Small	Small	Medium	Large
Product variations	Very high	High	High	Low

FIGURE 2.13
Overview of different entry points of customer orders. (Generated by authors.)

2.2.7.1 Assembly-to-Order

The assembly-to-order strategy is characterised by customers' ability to configure their own products from standardised components. In this environment, standard work can be implemented by modularising the work instructions according to the product structure. Processes, operating resources, technical parameters, and times can be assigned to each module in advance. Digital worker assistance systems can then be used to provide suitably configured instructions at each workstation and to make settings for the work system (e.g., for times, screwdriving sequences, torques, dosing, or material withdrawal).

2.2.7.2 Make-to-Order/Engineer-to-Order

In the make-to-order business, the individualisation of an existing product system is usually carried out through customisation development. This results in the customised production of components for the respective order. New drawings and part numbers are created. Tools and new materials may have to be procured, and programs and work instructions may have to be created. The same applies to engineer-to-order, i.e., developing new components and machines according to customer requirements. In extreme cases, engineer-to-order can also lead to unique manufacturing and assembly processes. However, the transfer processes from engineering to the shop floor are also repetitive. In addition, classes of similar products and components can be formed (depending on complexity, volume, labour content, etc.), to which standard times – derived from empirical values – can be assigned. These times can then be assigned to individual orders' production and assembly steps to create target capacity profiles according to the work progress. These can be compared with the actual progress and the confirmed times. This also enables deviation management on a daily basis in the engineer-to-order environment.

The different strategies – and, therefore, different standardisation approaches – can also occur in different places in the same company.

A manufacturer of large machine tools assembles them at a fixed station. Before assembly begins, the assembly planning department draws up plans for the assembly progress of individual machines based on time specifications for the components to be installed. This allows an assembly progress plan to be drawn up for the respective component configuration of a machine. Each assembly station has a terminal for ongoing feedback on the assembly progress. This feedback is used every morning to provide an overview of the actual progress compared to the plan for all locations. This means that deviations can also be identified in project business and countermeasures can be taken in a timely manner.

2.2.8 Maturity Levels in Dealing with Process Specifications

The previous chapters clarified what process specifications are and how standards and key figures can be successfully defined. Now it is a matter of using them correctly on a day-to-day basis. On the one hand, they must not disappear "in a drawer". On the other hand, they should not be misused to control employees. In both cases, the CIP will not be supported by the employees. In order to support the CIP appropriately with process specifications, care must be taken to ensure that they are handled appropriately (Figure 2.14).

First, process specifications are not valid forever. They only record the current best-known process and must continuously adapt to the environment and new findings. Process specifications help train new employees. However, they do not make employees interchangeable. In many cases, the skills and abilities of employees play a major role if process specifications are to be reliably adhered to. This applies in particular to small series production, as not every detail can be adequately described. In addition, employees' experience and process knowledge are explicitly required to develop processes further and adapt standards. Strict attention must be paid to separating the operational standard work from the improvement work within the framework of the DM. Otherwise, improvisation and unsecured activities will occur, posing a risk to product quality.

Last but not least, process specifications and the DM must not be used as a monitoring tool to identify and punish culprits. If deviations from target

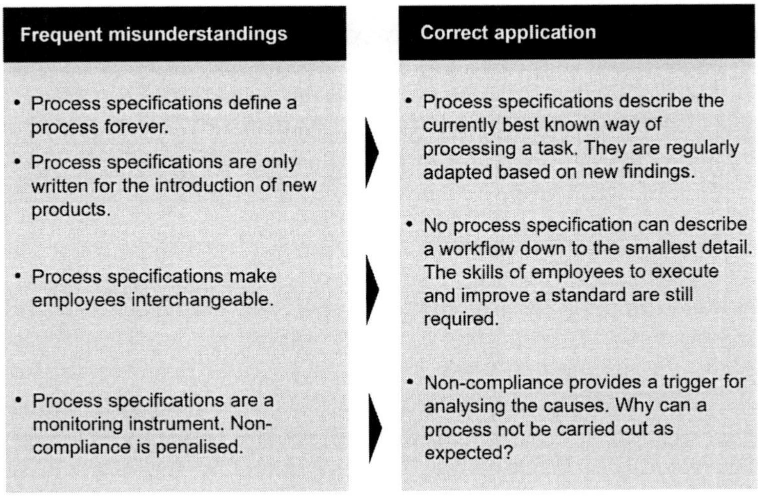

FIGURE 2.14
Misunderstandings and the correct application of process instructions. (Generated by authors.)

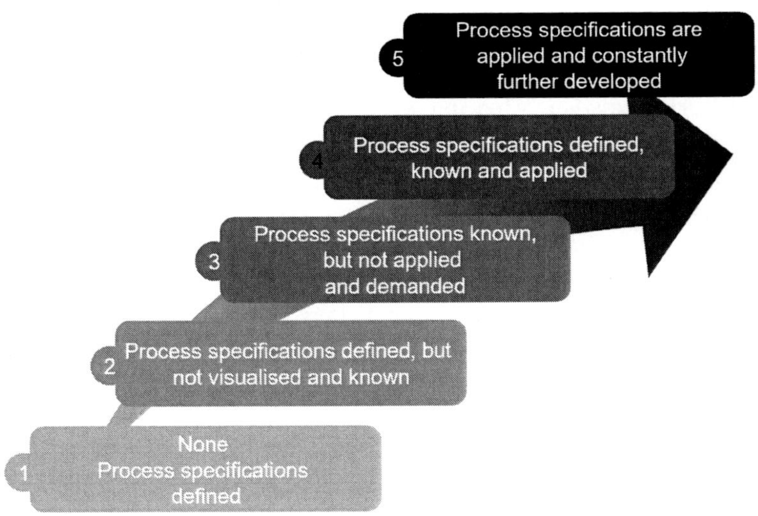

FIGURE 2.15
Maturity levels in dealing with process instructions. (Generated by authors.)

values or standard specifications are identified, the focus is on analysing the causes. The principle of improvement is: "The right process delivers the right result". Conversely, we initially assume that a process needs improvement if its result is not correct. That's why we first ask "Why?" and not "Who?" in the event of a deviation.

The five maturity levels for handling process instructions can help facilitate self-assessment in managing process instructions and the CIP based on them (Figure 2.15).

Process specifications arise from the proactive definition or improvement of a process. If they are not available, the basis for a target/actual comparison necessary for CIP is missing (1). If process specifications exist but are unavailable or known to employees, they cannot support a systematic CIP (2). Training and visualisation are the prerequisites for their daily application. However, even known standards and targets will not be implemented if they are not requested and demanded by leaders on a daily basis (3). New processes, in particular, sometimes require months of on-site presence by team leaders and managers before they are practised and implemented in daily routines (4). The highest level in terms of practised CIP is reached when process specifications are not only adhered to but also reviewed as part of deviation management (reactive improvement), enriched by knowledge, and continuously developed further (proactive improvement) (5).

The interaction of the improvement processes is explained below.

2.3 Proactive and Reactive Improvement

We essentially differentiate between proactive and reactive improvement. Where artificial intelligence is used, predictive improvement will also play a role in the future. This chapter examines how these types of improvement interact and what role standards play.

An improvement process usually starts with the initial standardisation (Figure 2.16).

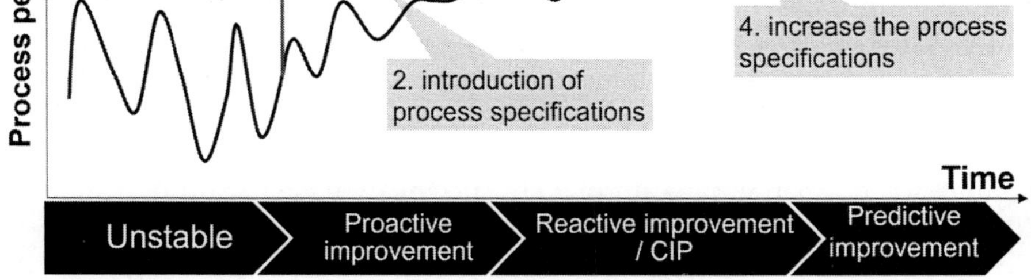

FIGURE 2.16
Relationship between process specifications and process performance in the phases of an improvement process. (Generated by authors.)

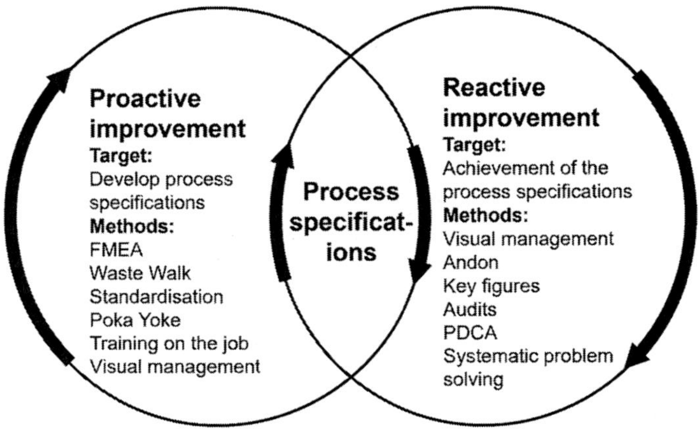

FIGURE 2.17
Interaction of proactive and reactive improvement. (Generated by authors.)

The unstable phase of a process is overcome by eliminating the main wastes and sources of variability (1). A start can be made on defining standards and setting targets. This completes the first phase of proactive improvement (2). The process enters the reactive improvement phase, where target and actual performance are continuously compared for selected key figures. Identified deviations trigger reactive improvement activities (3). If the targets can be achieved with process reliability or new strategic requirements arise, the process targets are continuously increased to achieve the desired improvements: proactive improvement through CIP (4).

With increasing data availability and suitable analysis methods, it is becoming increasingly possible to recognise causes and problems before they negatively impact the process result and thus become a deviation. At this point, the improvement process even enters a predictive phase: Recognising a cause of deviation at an early stage can, if necessary, enable countermeasures to be taken before a drop in performance occurs (5). Chapter 6 shows examples of this.

The interplay between proactive and reactive improvement is of interest to the CIP within the framework of the DM. They differ in their objective and in the methods used (Figure 2.17). It is essential for workshop managers to recognise which phase of improvement a process is in so that they can choose the appropriate improvement approach.

As part of proactive improvement, non-value-adding waste is eliminated so that it is not perpetuated in process specifications (e.g., through some of the basic lean methods, "spaghetti" diagrams, "stand in the circle", or workflow analyses). Analysing possible sources of error and avoiding them by means of a process – Failure Mode and Effects Analysis – can also improve the definition of process specifications. The specifications must then be recorded (standardisation) and implemented daily. It is essential to instruct

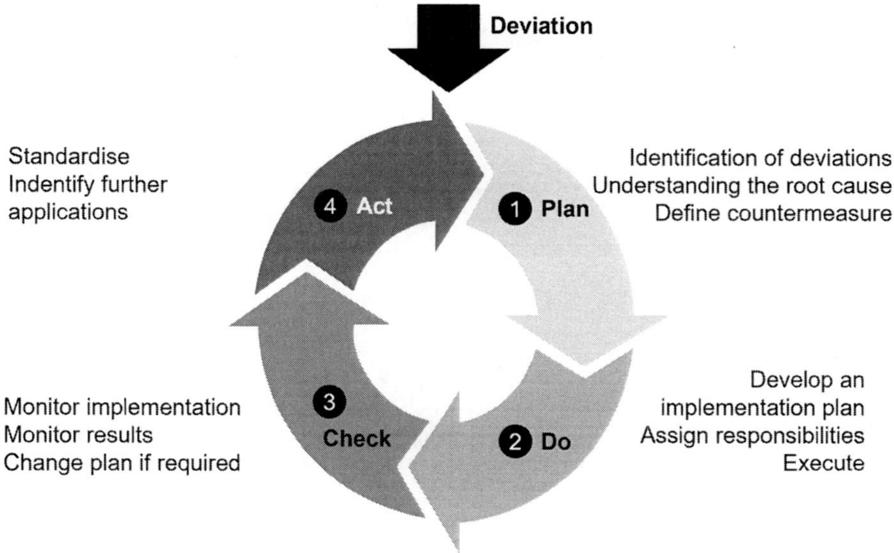

FIGURE 2.18
The PDCA cycle. (Generated by authors.)

employees on the new specifications and provide on-site support during the implementation phase. Enforcing the specifications with visual management and fail-safe design (poka-yoke) already leads to a reactive improvement. Key performance indicators, targets, and audits, further help to identify deviations and remedy them using the PDCA cycle.

2.4 The Flywheel of Improvement: PDCA

Whether proactive or reactive, improvements are only successful if they are anchored sustainably. The basic procedure for achieving this consists of four steps: planning (Plan), implementing (Do), checking (Check), and anchoring (Act) – commonly known as the PDCA cycle.

Plan-Do-Check-Act ensures that root causes for deviations are identified, appropriate countermeasures defined and reponsibilities assigned (plan), that these measures are implemented (do), their effectiveness is checked (act), and the new findings are consolidated in everyday life (act) (Figure 2.18) [3].

The individual steps can be processed with varying degrees of effort. Table 2.2 summarises the steps within the PDCA cycle.

Phase	Activities	Methods
PLAN	Forming a problem-solving team	Composition of the problem-solving team based on the competencies of those involved Interdisciplinary and cross-departmental team [4] Some companies have fixed teams with their own regular meetings
	Root cause analysis	- Create a specification for the problem - Collect possible root causes - Test possible root causes against the problem specification - Continue working with the root causes that pass the test
	Planning of measures	Finding a suitable, sustainable solution Selection of the best approach [5] Define responsibilities and resources Fix due dates Achieve mutual agreement for the implementation plan
DO	Realisation of immediate measures	Prevent the error from reaching the customer with immediate measures
	Realisation of the measure	- Track progress against the plan - Support the team and remove obstacles - Lead from the shop floor
CHECK	Review of effectiveness	Measuring success with defined criteria [6] It may be that the cause that has been eliminated does not turn out to be the correct one If necessary, repeat steps from the root cause analysis with new knowledge
ACT	Standardisation	Transferring the successful solution into standards to prevent this deviation in the future [7]
	Distribution of the acquired knowledge	Inform areas with similar processes Consolidate the acquired knowledge through wikis or training courses

This approach allows problems to be addressed at the root cause and be eliminated long term. However, it requires some effort and personal problem-solving skills. It is, therefore, the task of managers to convey and exemplify the skills and importance of problem-solving.

2.5 Management Culture in the Continuous Improvement Process

Much has been written about the right leadership culture in connection with lean transformations. Therefore, only the aspects relevant to leadership in DM and the CIP are outlined here.

From Optimising Numbers to Improving Processes

Achieving measurable goals is also a focus for managers in DM. However, they understand that financial results are the result of processes. The focus must, therefore, initially be on process design and improvement. The basic conviction is that better processes lead to better results.

From the Five "Who" to the Five "Why"
Good DM leaders do not jump to conclusions or jump straight to solutions. They first try to understand the situation on the ground and then ask, "Why?" Finding and "punishing culprits" is not part of systematic problem-solving and negatively affects the participants' motivation.

From Problem Hiding to Problem Solving
The DM assumes that not everything goes according to plan. Therefore, the mindset is: "No problem is a problem". This means that if no problems are reported from one area, no further improvement will occur, or problems will be concealed. The management must, therefore, demand that problems be reported instead of expecting only good news.

From Telling "How to Do" Things to Telling "What to Achieve"
Assigning challenges to individual employees as part of the DM is seen by managers as a way of encouraging people to take responsibility, solve problems, and expand their competencies through learning cycles.

From the Authority to Make Decisions to the Authority to Let Decisions Be Made
In DM, managers generally do not specify solutions in detail but formulate process specifications and target states. They then give employees the responsibility of proposing solutions to problems and driving their implementation themselves. This is part of the development philosophy.

From Giving Answers to Asking Questions
If a manager has a tendency to know better and gives the answer to every question about a problem, it prevents employees from independent thinking and taking responsibility. In DM, it is more important for the manager to ask the right question than to give the right answer. The rule is: He who asks, leads!

From Self-validation to the Validity of Others
The additional commitment of DM employees requires creativity and motivation. Even if not all activities are successful, successes will not be long in coming. These successes do not belong to the manager. It is additionally motivating to emphasise individuals or teams for success, for example, by presenting their CIP activities to the management.

This understanding of leadership to promote CIP by employees is implemented operationally in DM.

References

[1] Liker, J. K.: *The Toyota Way. 14 management principles of the world's most successful automotive group*. Munich: FinanzBuch-Verl. 2006.

[2] Liker, J. K.; Meier, D.: *The Toyota way fieldbook. A practical guide for implementing Toyota's 4Ps*. New York: McGraw-Hill 2006.

[3] Deming, W. E.: *Elementary principles of the statistical control of quality. Series of lectures*.Tokyo: Nippon Kagaku Gijutsu Remmei 1951.

[4] Pyzdek, T.: *The Six Sigma handbook. A complete guide for green belts, black belts, and managers at all levels*. New York: McGraw-Hill 2003.

[5] Sobek, D. K.; Smalley, A.: *Understanding A3 thinking. A critical component of Toyota's PDCA management system*. Boca Raton, Fla: CRC Press 2008.

[6] Garvin, D. A.: Building a learning organisation. *Harvard Business Review* 71 (1993) 4, pp. 78–91.

[7] Liker, J. K.: *Der Toyota-Weg. 14 Managementprinzipien des weltweit erfolgreichsten Automobilkonzerns*. München: FBV 2014.

3

Daily Management

ABSTRACT This chapter explains the concepts used in DM using the DM control loop and shows how the described ways of acting and thinking are interlinked. The characteristics of a successful DM are then illustrated through a practical example.

3.1 The Daily Management Control Loop

The proactive and reactive improvement processes must be supported, exemplified, and implemented daily. The DM control loop operationalises these approaches for everyday life.

DM is a method to manage the shop floor effectively [1]. It concerns the shop floor, not top management. DM aims to achieve operational goals and proactively improve organisational and technical processes, not to create reports. This also requires developing the problem-solving skills of employees and the organisation.

The DM control loop (Figure 3.1) describes the process and the interaction of the instruments and concepts within the DM framework [2].

The developed process specifications (see Section 2.2) are compared daily with the actual process performance to recognise deviations. Their causes are then analysed. If this is not possible in a short exchange and the deviation is important enough, a systematic problem-solving process (PSP) is started to find the cause. Measures are then taken to eliminate the causes according to the PDCA logic. The new findings are then fed back into the process specifications. The DM control loop is implemented in short, daily meetings across all management levels (Figure 3.2).

The meeting cascade begins at the team level. The actual performance is compared with the target, and deviations are recorded. If these cannot be dealt with within the team, they are escalated to the next level. Performance data, problems, and measures are communicated upwards, and the respective management feeds back decisions, information, and process objectives [1].

DOI: 10.1201/9781003612681-3

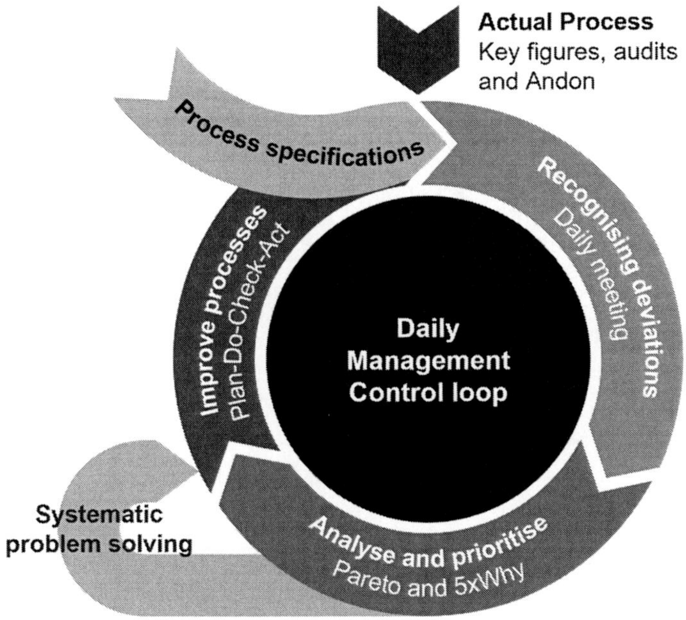

FIGURE 3.1
DM control loop. (Generated by authors.)

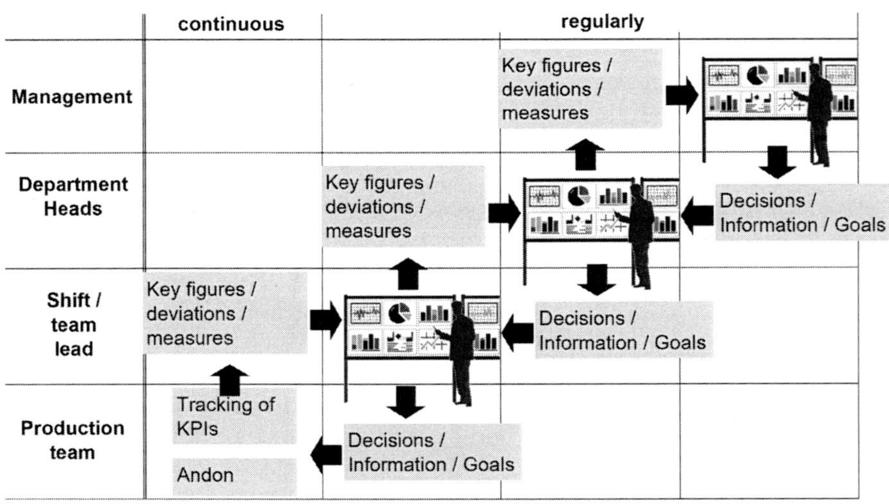

FIGURE 3.2
Meeting cascade in DM. (Generated by authors.)

In order for the DM control loop to function effectively in the meeting cascade,

- Employees must understand the process specifications and be able to monitor actual performance at all times.

- The actual process performance must be transparent.
- Effective deviation management in the sense of PDCA must be anchored in the organisation.
- The problem-solving skills of managers and employees must be developed.
- Managers must demonstrate a disciplined approach to systematic problem-solving.
- Clear handover points must be defined between the stages of the cascade and times, and participation must be disciplined.

The creation of process instructions is explained in Section 2.2. The subsequent steps in the DM control loop are explained operationally below.

3.2 Making the Actual Process Transparent

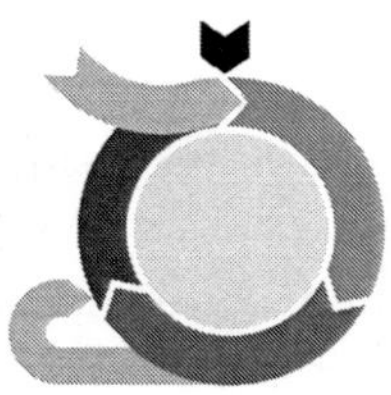

In order to recognise whether process specifications have been achieved, process performance can be made transparent during the daily meeting using key figures. Andon and Gemba Walks and audits are other tools that help to recognise deviations directly on-site.

3.2.1 Andon

Andon describes the shortest response cycle to ensure that shift targets are met. Employees or machines and systems immediately draw attention to themselves in the event of deviations and ask for support (Figure 3.3).

A typical Andon process runs as follows: A deviation occurs at a workstation. The employee requests support by signalling ("yellow" signal). The team or shift leader can usually resolve the deviation. If this is not possible within the remaining time of a cycle, the entire line is stopped ("red" signal). As a rule, the production process is not continued until the problem has been resolved. This will

- Drastically reduce the time from deviation detection to reaction,
- Analyse a problem as soon as it is discovered while it is still "fresh",

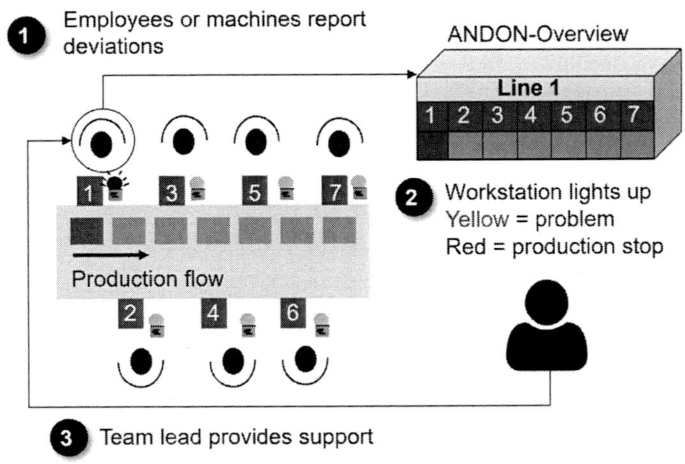

FIGURE 3.3
The Andon process. (Generated by authors.)

- Reduce the risk of errors being passed on to the production process,
- Sensitise all those involved to a source of error, and
- Ensure that the problem-solving process is well-prepared through a thorough specification of the problem immediately after discovery.

Escalation rules are often defined: for example, production management is informed after 15 minutes of downtime, and management after 60 minutes. Andon can be very well supported digitally: for example, managers are warned online or called in.

ANDON is suitable for quickly pointing out deviations before too much downtime occurs, capacity is lost or waste is generated. Every ANDON incident should result in an entry in the action plan and a vote by the DM team on whether the DM cycle should be continued.

3.2.2 Gemba Walks/Audits

Gemba walks (from the Japanese "genba" = "place of value creation") are standardised tours of a production facility in which the status of production, assembly, testing processes, logistics, etc., is routinely checked using defined reporting points. As a rule, managers receive a short, standardised report from the employees at each point. The Gemba Walk can cover both direct (production) and indirect areas (Figure 3.4).

In production, for example, it is worth looking at the delivery performance, the quality situation, and the status of ongoing improvement activities. A "deep dive" is a good idea for individuals who are changing workplaces daily, and it includes a discussion with the on-site employees. In this

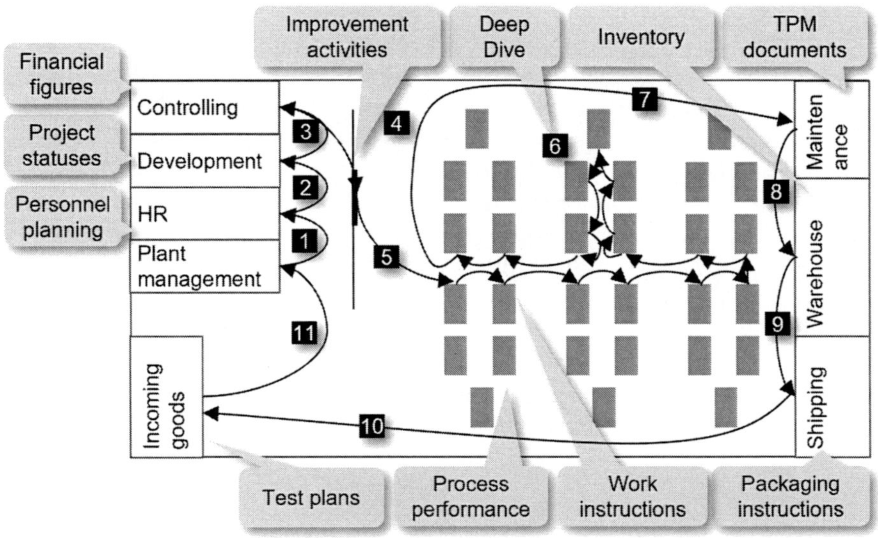

FIGURE 3.4
Example of a Gemba Walk. (Generated by authors.)

discussion it will turn out whether associates know and understand their goals and targets, how problems are dealt with, and what should be further improved can be recognised.

This element is often practised in weekly and monthly 5S audits. However, a shorter daily rhythm and reviewing all types of process specifications in all areas are more effective, as they connect leaders closer to what is happening at the shop floor and employees feel appreciated and interested in their work. In order to manage the Gemba Walks at such a high frequency, it is helpful to focus on different small areas every day.

3.2.3 Key Figures

The actual values for the defined key figures (see Section 2.2) are recorded daily, preferably by the team itself. Traditionally, the employees mention outputs, quality problems, or accidents in the meeting. Today, much of the information can already be found in the ERP or MES and retrieved there. How and by whom the daily data collection process is carried out should be part of the KPI definition so that comparable figures are available daily.

During a digitalisation project at a sensor manufacturer, data from the MES was to be used for DM on the shop floor. The central lean team defined the key figures that were to be valid across different production sites. Once the MES was connected, it became apparent that the key figures did not match those that the local teams had on their analogue boards. While the reams pulled the reports from the system, they evaluated different product groups differently; for example, they defined the completion of a process using different reporting points to reflect the reality on-site better. However, these changes in the data collection process were never formally recorded or carried out consistently across the teams or over a longer period of time. The comparability of the values was thus lost, and the data collection process had to be redefined for most of the key figures before the dDM project could be continued.

The data is collected in preparation for the daily meeting, where it is compared with target and limit values to recognise deviations.

3.3 Conduct Daily Meeting

The meetings on the DM board are the most striking element of the DM method, as this is where the whole team meets. The success of the meetings and, therefore, the method depends heavily on the moderation of the meeting.

3.3.1 The Daily Management Board

The central element of DM is traditionally a whiteboard in production. Key figures, their trends, the status of measures and problem-solving processes, general information, the agenda for the meeting, and rules of conduct are visualised on this board (Figure 3.5) [3].

The board should be designed so that a person from outside the department can immediately recognise the status of the production area – both with regard to the achievement of results and the ongoing measures. The DM board is used for regular meetings.

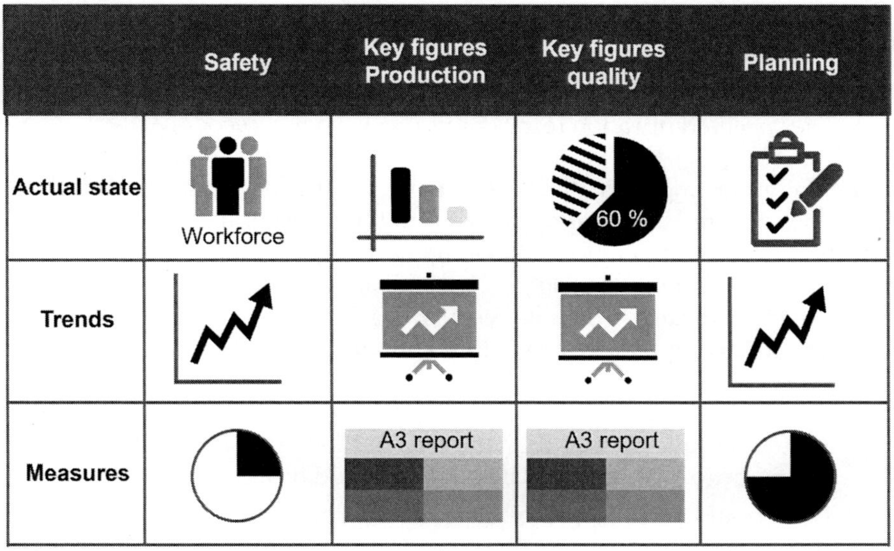

FIGURE 3.5
Analogue daily management board (schematic). (Generated by authors.)

3.3.2 Moderation of the Daily Meeting

The direct manager should moderate the regular DM meetings – this creates commitment. It makes sense, for example, for the head of the department to moderate the meetings of a group for an initial period and then hand over the moderation to the group leader. For the subsequent meetings, the head of the department stays in the background and gives feedback to the group leader after each meeting until they have mastered the new leadership routine. A substitution arrangement for the moderator should be defined to prevent the meeting from being cancelled due to the absence of a team or group leader. In order to qualify additional people for the moderation, a rolling moderation makes sense. It is not practical to have the DM meeting conducted by lean managers, as they generally have no disciplinary authority and are only available for a limited period of time.

For a successful DM meeting, the moderator must ensure the following points:

- All participants are on time.
- All participants have prepared their information.
- A standard agenda is processed.
- Deviations and problems are addressed openly.

- Suggested solutions are requested from the team.
- If causes and measures cannot be clarified directly, the problem solution is outsourced from the meeting.
- The specified duration (e.g., 15 minutes) is not exceeded.

In the daily meeting, it is a major challenge for the moderator to collect the deviations and the information relevant to their description. Unfortunately, speeches are often made – whether intentionally or unintentionally – that are based on opinions or distract from the actual topic. It is important for the moderator to distinguish between what is relevant and what is irrelevant. The following pattern can help the moderator prevent a meeting from digressing:

- Firstly, the question of the target or normal state in which a process should be (e.g., due to specifications) should be asked.
- This should be followed by the question of the observed actual state and in which categories, figures, or behaviours it deviates from the target state. As a general rule, only facts are permitted to describe a deviation.
- Only then is a brief discussion about possible causes and measures desirable and sensible. A countermeasure can be defined immediately during the meeting if the cause–effect relationship is clear. However, a discussion about causes goes beyond the time frame and usually does not involve everyone present. The moderator must collect the relevant information but then stop the discussion and move the problem-solving to another time and possibly to another team.

This allows the moderator to find the right balance between pure key figure reporting and a discussion round – because the employees quickly reject both.

If only figures are exchanged in the meeting and, in the worst case, these are underlaid with accusations, the meeting may be short, but no improvements are initiated. If the team is neither asked for solutions nor offered any, it (rightfully) quickly loses interest.

The other extreme, which leads to rejection, is drifting off into discussions. This leads to overrunning meetings, and usually, not everyone present has an opinion on a topic – if opinions rather than facts dominate the discussion, it is ineffective. In order to keep the meetings efficient for everyone, the moderator must outsource the discussion and take measures to obtain new facts, for example, by bringing in other experts or presenting an emotionally charged topic objectively using their own key figures.

You can often tell from the posture of the participants whether a DM round is successful. If a large proportion of them are standing with their arms folded, leaning casually, or just looking at the exit door, they are making it clear that they don't see any point in the meeting.

At the plant management level of a clutch manufacturer, the dynamics of the DM meetings changed from one day to the next due to different moderations. The plant manager focused strongly on problem-solving and, above all, wanted to know what was currently being worked on and how he needed to provide support on the individual topics. Deviations were addressed, and decisions were made in the meetings. On Tuesdays and Thursdays, a deputy took over the moderation and went through the same agenda items. However, this person was essentially informed about incidents, and the meetings had far fewer productive results. After external observations uncovered this, the plant manager led the DM meetings on all days, and the deputy was again trained in the importance and objectives of DM.

3.4 Analyse and Prioritise Deviations

The organisation must develop two routines for evaluating and processing the recorded deviations: Create complete deviation statements and prioritise them clearly.

3.4.1 Analyse Deviations

Successful deviation management and any necessary problem-solving process begin with describing deviations completely and independently of the solution. The description of the deviation ensures a uniform understanding of the problem. It is thus the foundation for the subsequent root cause analysis [4]. An inadequate description of a deviation may lead to

- Inadequate understanding of the problem
- Solving the wrong problem
- Searching for the wrong cause
- Reworking the process because the problem occurs again
- Ultimately, the frustration of the DM team

A good deviation statement gives the deviation a precise title. The deviation should be described as the difference between the target and actual state and should always include the affected object. For example:

> Robot 1 of the brake pad assembly line positions the tenth segment incorrectly throughout.

The object of the deviation is the robot of the brake pad assembly line, the deviation is the positioning at the wrong location.

In addition, a good deviation statement points directly to where the error occurred because this is where the problem-solving process should start. An example of a weak deviation statement would be:

> Assembly line A produced 20 parts below target in the morning shift.

In order to obtain a better, more specific description of the deviation, the moderator of the DM round should now try to get closer to the source of the error by asking "Why?" questions:

"Why was the A line 20 below target?"
└> Answer: "Because the employee at the third station couldn't follow the beat."
"Why was this the case?"
└> Answer: "We don't know yet."

The deviation statement is now as follows:

> The employee at station 3 of line A cannot follow the cycle time continuously.

This deviation statement is much more specific than the first one and points directly to where the search for the cause should start.

A good deviation statement is, therefore, of fundamental importance for two reasons:

- It provides the exact title for further communication about the problem and gives the deviation a name.
- It provides the framework for prioritisation and, if necessary, further search for the cause.

3.4.2 Prioritisation of Deviations

If every deviation is treated with the same priority, the problem-solving capacities of a team are quickly overtaxed. The quality of the problem solutions will suffer, and frustration can result as issues remain unresolved for too long.

Therefore, one of three action alternatives must be selected for each deviation detected (Figure 3.6).

Either deviations are only discussed, with no activities initiated, measures taken, or problem-solving efforts started. This decision depends on two characteristics: Is the issue important, and is a cause already known?

Deviations that are not currently important do not need to be processed directly. Priority can be recognised based on target values or frequencies.

If the deviation is to be addressed, the next question is "How?" If the team can identify the root cause, measures can be defined directly. Otherwise, a problem-solving process is required in advance to analyse the cause.

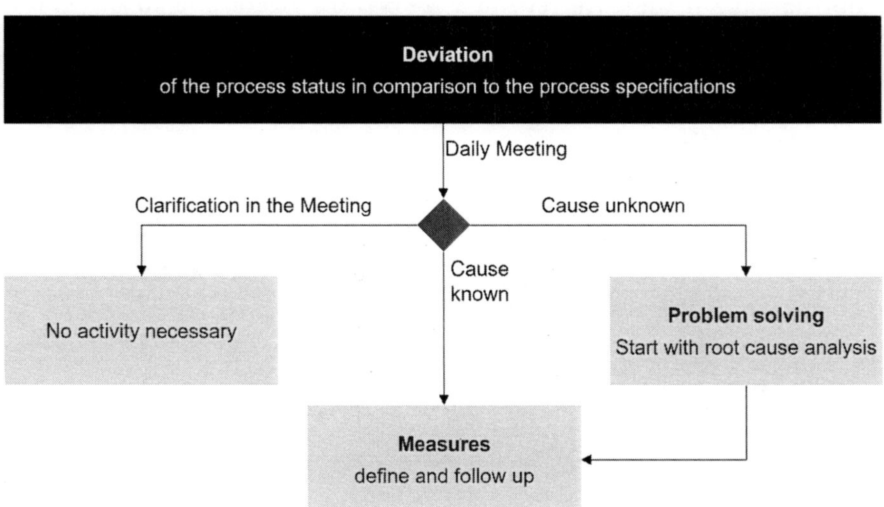

FIGURE 3.6
Possible paths of a deviation. (Generated by authors.)

3.4.2.1 Prioritisation Based on the Target Values

The first step in differentiating between the important and the unimportant is to focus on the target values of a process. The first question to be answered is whether addressing the existing deviation or solving a problem brings the process closer to this goal (Figure 3.7).

Assume that the scrap rate should be a maximum of 3% (target status) and is currently at 6% (actual status). The employees repeatedly report problems or make improvement suggestions. It is then necessary to prioritise: Which problem directly prevents achieving the target status? For example, excessive setup times or inadequate material supply would be issues that could be addressed, but this does not directly reduce the scrap rate. In this example, machine parameters or employee qualification should be addressed one after the other through several PDCA cycles to achieve the target state (black flashes in the dark grey corridor, see Figure 3.7).

In this case, deviations from the specified setup times would not currently be addressed, as quality issues take priority. Nevertheless, it is important to continue recording the deviations to prioritise them later based on their frequency.

3.4.2.2 Prioritisation Based on Frequency

Even deviations that seem small and unimportant on their own can cause major damage if they are repeated frequently. According to the Pareto principle, 80% of performance losses in a process can be attributed to

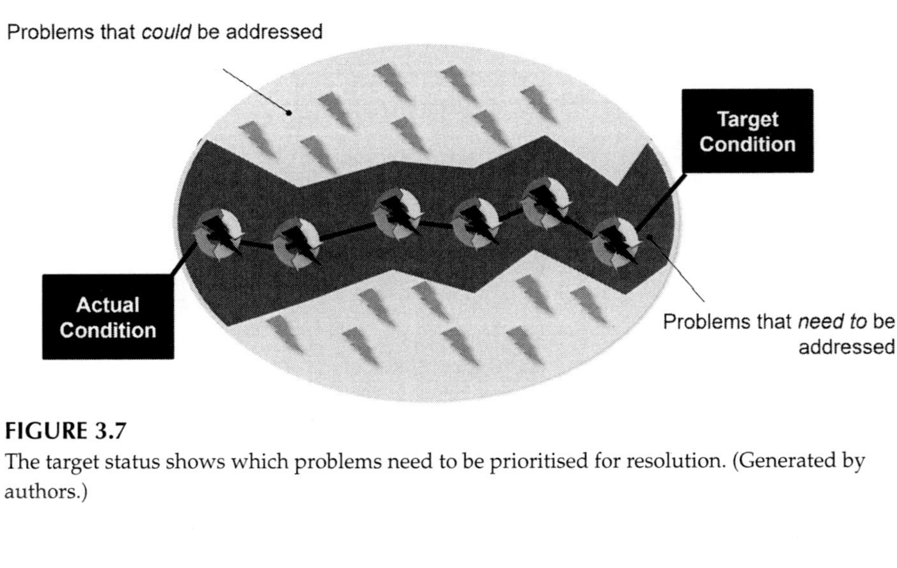

FIGURE 3.7
The target status shows which problems need to be prioritised for resolution. (Generated by authors.)

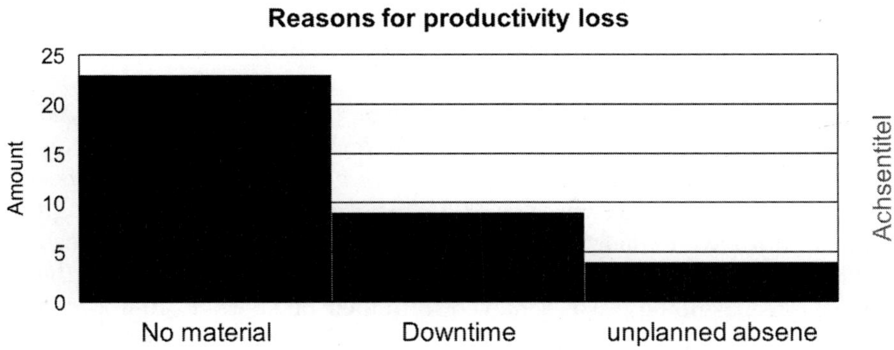

FIGURE 3.8
Example of a Pareto chart. (Generated by authors.)

20% of the causes [5]. Finding and eliminating that 20% can significantly improve the company. In practice, this is realised through Pareto analyses (Figure 3.8).

In this analysis, the deviations are categorised in order to identify the most frequent causes and problem areas. The frequencies of deviations, the sums of disruption durations, or follow-up costs can be used to scale the analyses [3]. In dDM, the most frequent deviations can also be determined directly by analysing the full texts (Section 7.3).

If a deviation is recognised as important, measures must be defined, but this can only be successful if the root cause is known. But how is a root cause defined or recognised?

The term "root cause" refers to the cause that started the chain of effects that led to a deviation. Therefore, the first objective of measures against a deviation should always be to eliminate this root cause. Once the root cause has been reliably and sustainably addressed, the associated deviation can no longer occur. If the root cause of a deviation is already known to the DM team, the search for suitable countermeasures can begin immediately, and if necessary, a measure can be defined during the meeting. As a rule, this applies to the majority of recorded deviations.

To recognise whether the root cause has already been identified, the team can ask itself: "Will eliminating this cause reliably prevent the deviation from occurring?" If the answer is "Yes", the root cause has been found, and no further problem-solving is necessary. In this case, the team can immediately define a countermeasure. This measure is then followed up according to the PDCA logic on the team action plan (Section 3.5).

If the root cause cannot be identified in the DM meeting, a problem-solving process should be started, which is not part of the DM meeting due to time constraints (Section 3.6).

A large car manufacturer wanted to overhaul its deviation management in engine assembly. Structured problem-solving processes and responsible teams were already in place. However, the action plan comprised several hundred open points. As prioritisation with "1", "2", and "3" was no longer sufficient, a "0.9" and a "0.8" were introduced. This clarifies that even problems prioritised as "1" were no longer processed. Although great efforts were made to solve problems and many issues were dealt with, the number of unresolved issues still led to frustration among the workforce, as nothing was achieved from their perspective. To improve the situation, the existing action plan was initially cancelled – some of the content was outdated, and the backlog was no longer manageable. To prevent this from happening again, the criteria for starting a PSP and for escalation were made transparent. The first and most important criterium for prioritising an improvement opportunity is the question: "How much will solving this issue contribute to achieving our targets for the assembly line? Additionally the reponsibility for prioritising issues was transferred to the team leaders and their teams. This avoids overloading the higher levels with to many tasks while increasing employee satisfaction.

In the example of engine assembly, the three deviations with the highest priority and the three most frequent are selected each week for systematic problem-solving. Six problems per week are already a lot, and there are additional regular meetings to process the problems with maintenance and production. Without additional capacity, one to two systematic problem-solving processes per month is a good target.

3.5 Implement Process Improvements

Once measures have been defined, they must be implemented quickly. To this end, the status of ongoing measures is tracked on the team action plan and escalated if necessary.

Date	Deviation	Measure	Responsi ble	Dead line	PDCA Status
23.5	Machine standstill due to overheating	Shorten lubrication cycles	Marvin Müller	4.6.	
27.5.	Defective parts were passed on	Provide red containers for defective parts	Christian Hertle	7.6.	
...					

FIGURE 3.9
Team action plan (example). (Generated by authors.)

3.5.1 Team Action Plan

The team action plan (Figure 3.9) is the central document where deviations are recorded, briefly described, and assigned measures. A look at the timeliness and care with which a team action plan is kept shows whether the associated CIP is a living process.

On the team action plan:

- Deviations are recorded, and deviation statements are documented.
- New measures are defined.
- Responsibilities are for measures assigned and scheduled.

In addition, the progress of measures from previous meetings is tracked in the team action plan according to the PDCA logic.

"Harvey balls" (circles with four segments) are used to assess the status of implementing activities. As part of the approach favoured here, these are not – as is often the case – used to track the percentage progress of the implementation of measures. Rather, the progress of measures should be tracked according to the four steps of the PDCA logic:

1. Plan: Measures defined and implementation planned
2. Do: Measures implemented on-site
3. Check: Measures are proven effective
4. Act: Measures transferred to the new standard and applied to other processes.

The activity is only closed after the fourth step.

The DM meeting is the time to review the activities already defined based on the team action plan. This means reviewing the team action plan and having those responsible report on the status of outstanding issues. Progress can be reviewed, possible obstacles discussed, further support measures defined (e.g., by the manager) – or measures escalated.

3.5.2 Clearly Defined Escalation Paths

As far as possible, deviations should be dealt with directly where they occur. On the one hand, the best process knowledge is available in the teams; on the other hand, higher-level departments are not overloaded. However, this is not always successful. Often, the cause of a deviation lies in an upstream process, or the team lacks the decision-making authority to implement the measures. A cascade of meetings is therefore necessary to effectively distribute responsibilities and measures within the company.

> The meeting cascade must be used to establish a commitment to all areas for processing measures.

Time and again, the development department, for example, takes itself out of the cascade and the processing of measures. This impairs the organisation's learning ability and leads to frustration, as even simple solutions that require design changes are not implemented. There should also be a binding escalation channel to the development department, at the latest, via the management and preferably directly to the teams.

This is all the more important the earlier the entry point for individual customer orders is. In a make-to-stock process, products are usually tested and designed for series production. Therefore, the processing of acute measures by the development department is rarely necessary and a regular participation of the development department in the DM meetings is not required. In the ETO area, on the other hand, the specification of new products is often inadequate or incorrect. An exchange between production employees and development should occur at the team level. The same applies to purchasing and work preparation.

Escalation is important but should only be used in a targeted manner. If too much escalation is used, the problem-solving process becomes "clogged", and a few people in the company receive many measures and problems from different areas. The capacity for solving individual issues decreases, and the processing time increases. Such bottlenecks can be avoided by using escalation rules.

Escalation should take place (for example) if

- An end date has been exceeded by more than 5 days.
- An end date has been postponed more than 3 times.
- A measure costs more than €3,000.
- A measure has to be processed in another area.
- Three measures were not successful.

In addition, a limit can be placed on how many issues a team can escalate simultaneously (e.g., 3). This puts the responsibility for prioritisation in the hands of the on-site teams. They can best assess what is most important to them, and it is less likely that there will be no feedback on escalated issues.

Escalation along the meeting cascade also ensures that the forwarded measure is always a work order from the respective leader. The necessary commitment is lacking if measures are distributed across the company (e.g., via a digital system).

3.6 Systematic Problem-Solving

Measures cannot be defined directly for all deviations. If the root cause is unknown, it must first be identified. The tool for this is systematic problem-solving.

There are a number of established methods for analysing problems and subsequently finding their causes, on which excellent literature has already been published [6]. In addition, problem-solving does not occur during the meetings, as this would exceed their time frame. For this reason, only the basic methods and procedures for problem-solving in the context of DM will be discussed here: the detailing of the problem specification and the root cause analysis using 5xWhy or the Ishikawa method.

3.6.1 Kepner–Tregoe Method for Specifying a Problem

The point of cause is the time and place where the observed deviation is generated. It is the physical place to go to in order to collect more information to determine the root cause. The challenge for a problem-solving team is to get as close as possible to this point to start the root cause analysis.

In some cases, the point at which the error occurred cannot be clearly localised using the deviation statement, and additional information must be collected. Kepner–Tregoe method offers a helpful structure for this process of collecting and structuring information. The Kepner–Tregoe problem specification contains a detailed description of the problem in four dimensions:

- "What?" Type of deviation
- "Where?" Both the location in the hall and the exact location of machines or products
- "When?" and that includes "How often?"
- "How strong?" Extent of the deviation

All available information about the problem falls into one of these four dimensions. Within each dimension, questions are asked to describe which characteristics describe a deviation or how it presents itself to our senses. Facts must replace assumptions. If the facts are unavailable, they must be obtained, even if this takes time. The approach is "Genchi Genbutsu", that is, collecting the problem characteristics at the scene of the incident.

Here is an example in which the location of the origin of a deviation is still unknown. The deviation statement reads:

> Assembly line A has delivered 20 units below target defined by standard work specifications.

With the help of a table containing pre-formulated questions in each of the four dimensions, the problem-solving team now collects the characteristics of the deviation, which in the example leads to the following table:

Dimension		**Deviation**
What?	At which object is the deviation to be observed?	Line A
	What is the deviation precisely?	Output 20 pieces below target
Where?	Where do we observe the deviation (geographically)	End of line
	Where is the deviation at the object?	Station 3 doesn't keep up with the pace
When?	When was the deviation observed first?	This Wednesday in the morning shift
	When was the deviation observed since then? Is there a pattern?	Every shift
Extent	How many objects are affected?	Only line A is affected
	How big is each deviation?	20 pieces, every time

With the help of this collection of problem-related information, a more detailed problem specification can be formulated in addition to the deviation statement. In the example, it would read as follows:

> Assembly line A has achieved an output that is consistently 20 units below target every shift since Wednesday morning. Station 3 is unable to keep up with the customer takt.

Based on this problem specification, station 3 can be identified as where the search for the cause begins. This can be done using the 5xWhy method.

In cases where the point of cause cannot be identified in this way, the so-called logical comparison helps. This involves asking what could reasonably be a property of the deviation (=IS) but is not (=IS NOT). This provides a relevant basis for comparison for each property of a deviation, which allows additional conclusions to be drawn about the place where the deviation originated.

Dimension		**Deviation (IS)**	**Deviation (IS NOT)**
What?	At which object is the deviation to be observed?	Line A	Not line B or C
	What is the deviation precisely?	Output 20 pieces below target	
Where?	Where do we observe the deviation (geographically)	End of line	
	Where is the deviation at the object?	Station 3 doesn't keep up with the pace	Stations 1,2,4–7
When?	When was the deviation observed first?	This Wednesday in the morning shift	Since start of production
	When was the deviation observed since then? Is there a pattern?	Every shift	Limited to certain shifts
Extent	How many objects are affected?	Only line A is affected	
	How big is each deviation?	20 pieces, every time	Varying or upward/ downward trend

With the help of this table, the problem specification can be formulated even more precisely:

> Since Wednesday morning (and not before), assembly line A has achieved an output that is consistently 20 units below target. Only station 3 is unable to keep up with the customer takt. All shifts are equally affected. No other line and no other stations are affected.

This information clearly points to Station 3 of Line A as the location for troubleshooting and excludes other locations. It can now also be used to test the validity of suggestions for possible causes. The question of an experienced problem-solving facilitator could be: "If this is indeed the cause of the deviation, does this cause fulfil the problem specification in all dimensions?"

One example: a member of the problem-solving team suspects that the cause of the deviation is that a new employee is being deployed on station 3. This employee is working the early shift, has been familiarising themselves with the system since Wednesday morning and is therefore not yet able to follow the line takt. This suggestion for the cause can already be ruled out – without further analysis – because the deviation occurs consistently on all shifts and is not just limited to the early shift.

Problem-solving will be continued with the root cause analysis.

3.6.2 Problem-Solving with the Help of 5xWhy?

With the help of "5xWhy?" the root cause is identified by repeatedly asking about the reason for the deviation (reference value 5x) from the specification:

> Since Wednesday morning (and not before), assembly line A has achieved an output that is consistently 20 units below target. Only station 3 is unable to keep up with the customer takt. All shifts are equally affected. No other line and no other stations are affected.

Why?
↳ Because the employee often has to leave the station to fetch materials.
Why?
↳ Because the material delivery to station 3 is incomplete.
Why?
↳ Because incomplete part sets are created during order picking.
Why?
↳ Because the picking list was not adjusted after a change to the parts list.
Why?
↳ Because there is no workflow for automatically adjusting the pick list after the Bill of Material changes.

The answer to the last "why" question provides the root cause. A measure that automatically adjusts the pick list every time the parts list is changed will certainly prevent a recurrence of the deviation in question.

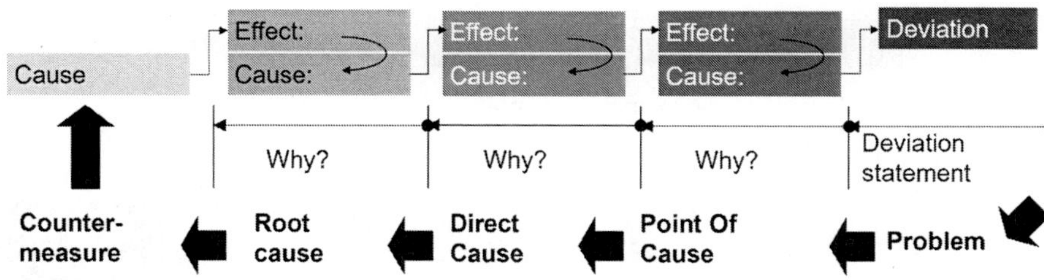

FIGURE 3.10
Analysing the cause-and-effect chain using "Why?" questions. (Generated by authors.)

The 5xWhy method is a simple and effective tool for analysing cause-and-effect chains. The key to its successful application is to trace the causal chain backwards from the observed deviation to a root cause step by step without losing the substantive link to the deviation (Figure 3.10).

The 5xWhy method initially appears to be a simple process. However, its successful application requires some practice. When the question "Why?" is asked, the person asking must assess whether the answer can be a possible cause for the effect mentioned. Therefore, the challenge is to stay on the causal path of cause-and-effect and not be led astray by obfuscating or unobjective answers. In addition, many deviations can lead to more than one causal path. The cause-and-effect chain then splits into several paths. This is where the Ishikawa method comes into play.

3.6.3 Ishikawa Method for Complex Cause-and-Effect Chains

If there are many possible starting points for the deviation or the cause-and-effect chain is highly branched, the problem-solving quickly becomes confusing. This is where the creation of an Ishikawa diagram comes in handy. Ishikawa diagrams systematically link the observed deviation with possible causes and provide an excellent overview of the structure of a problem-solving team's thought process. In order to structure this thought process, the root cause analysis usually takes place in five categories: Personnel, Method, Machine, Material, and Environment. In each category, the team asks questions in the following way: What causes could lead to the observed deviation in category X? In the aforementioned example of station 3 of assembly line A, the question in the "human" category could be as follows:

> If the employee is the cause of the low output, what are the possible causes?

Here, the 5x Why method is used to work through the possible causes to the root causes systematically. In the example, the root cause could be a lack of familiarisation process for new employees (Figure 3.11).

The question in the "Method" dimension could be:

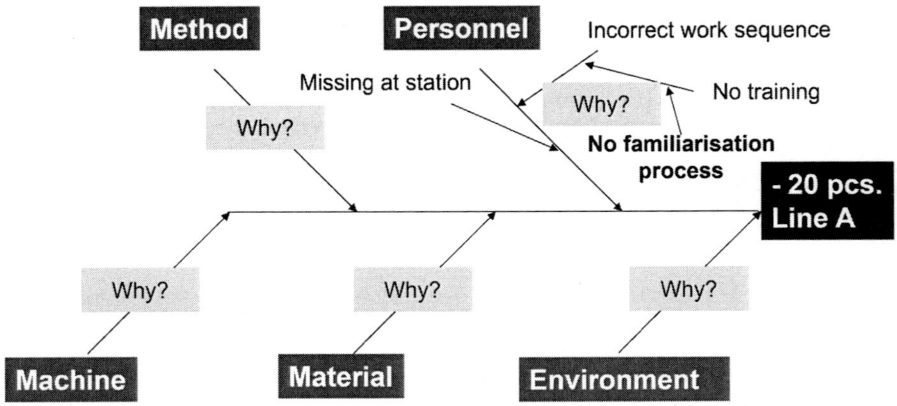

FIGURE 3.11
Possible causes in the "People" category in an Ishikawa diagram. (Generated by authors.)

> If the working method is the cause of the low output, what are the possible causes?

The corresponding why-question technique in the example means that the pick list in picking is not automatically adjusted when the Bill of Material changes (Figure 3.12).

The result of the Ishikawa method is, therefore, a series of possible causes for the deviation under consideration. These possible causes can, in turn, be tested to see whether they fulfil the previously defined problem specification in each dimension (IS/IS-NOT).

If so, measures can be defined and tracked in the team action plan according to the PDCA logic. If the check phase shows that the problem has not been solved, a further iteration of the search for the cause is necessary.

The methods for systematic problem-solving are simple in principle. However, most employees are not used to this approach and thinking as it is rarely required of them. The problem-solving skills in the teams, therefore, usually need to be specifically developed.

3.7 Competency Development

The moderation of daily meetings and, in particular, cause-centred systematic problem-solving in the teams require further development of employees. This is achieved by involving, challenging, and supporting them. Competent employees at all levels who can also deal with new challenges strengthen

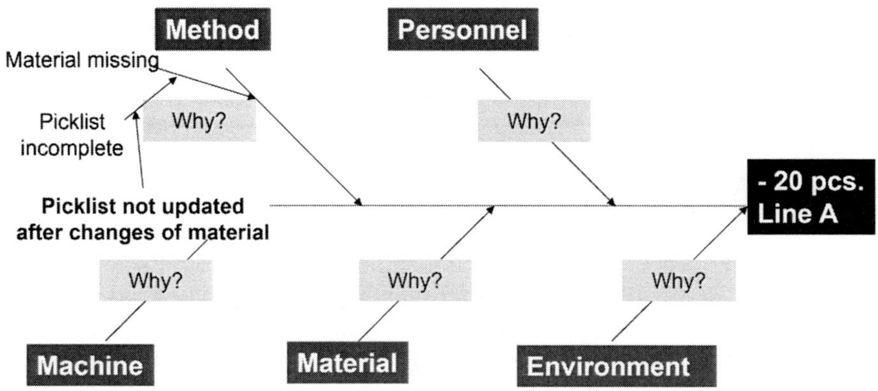

FIGURE 3.12
Possible causes in the "Method" category in an Ishikawa diagram. (Generated by authors.)

the resilience of their organisation. Especially in times of change, the topic of coaching is once again coming to the fore. This chapter explains the basic principles of this management approach.

Coaching is a management approach in which employees are not given direct instructions on how to solve a current problem but are supported in finding their own solutions by asking questions. This can lead to a different solution than the one that the manager would have developed. However, the employee gains confidence in solving problems independently and can take on more difficult tasks alone in the long term. Coaching requires patience and trust on both sides. The coach must not directly state the solutions they have in mind, and the employee should receive fewer direct work assignments. Instead, leading questions should be asked first and foremost [7].

The right questioning technique is an essential skill that managers need to learn. The aim is to guide employees through the systematic problem-solving process by asking suitable questions. Open questions are primarily asked to determine the direction but not the results. Closed questions should only be used to confirm understanding [6].

Here is an example of how, in the event of increased rejects on a machine, the "point of cause" is narrowed down by a questioner:

> "Why do we have so much scrap on machine no. 1?"
>
> "Because the fixture does not clamp the part securely, and it slips out of the fixture."
>
> "Why doesn't the device clamp the parts securely?"
>
> "Because the clamping pressure is below the specified value."

"Why is the clamping pressure below the specification?"

"We can't say at the moment."

"Did I understand that correctly? The clamping pressure is below the specification, which leads to unsafe clamping and scrap?"

"Yes, that's all we know at the moment."

The "point of cause" is reached when no further reason can be given in response to a (final) why question. In this case, the interviewee does not know of any further reason. Problem-solving, therefore, starts here. In the example, the next question could be if the clamping pressure is below the target value, what could be the possible causes? An Ishikawa diagram (Figure 3.10) can then structure the further questioning process by asking about causes in the area of operation, the machine, the material, etc. Figure 3.13 shows which questions a manager can use to guide an employee through the various phases.

Firstly, the employee should record the deviation themselves – and state exactly what deviates from a target and how. This involves asking whether targets have not been achieved and why. The next step is to analyse the

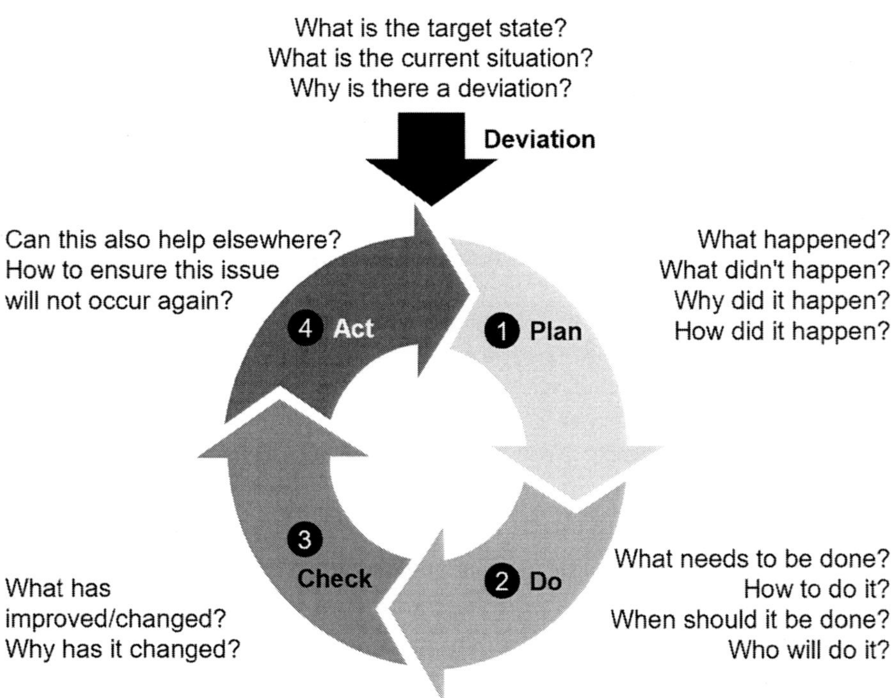

FIGURE 3.13
Coaching questions for the PDCA process. (Generated by authors.)

causes and discuss an action plan. After implementation, it is discussed what changes have been made and why. Finally, questions are asked about how the solution can be implemented in the long term.

Managers who are good at asking questions understand:

- The intention behind their questions.
- The assumptions they made when formulating a question.
- How important a careful choice of words is.
- How they will get answers.

In particular, team members who were previously managed hierarchically and whose independent thinking was suppressed rather than encouraged can quickly feel overwhelmed by this approach. However, it can be frustrating for more experienced problem solvers not to receive specific work instructions when stuck. Coaching is effective when it keeps employees in the "flow". This means the tasks are challenging but can be solved (Figure 3.14).

Every employee has a unique set of skills, be it technical, communicative, or problem-solving. This is why different employees evaluate the same task differently. Tasks that are too easy lead to boredom and frustration, and while long-term satisfaction can be found in the comfort zone, no development takes place. This is stimulated in the stretch zone, that is, a slight overload. If independent solutions are developed and implemented here, it will lead to a sense of achievement and satisfaction – flow. However, if the excessive demands last too long or are too strong, it leads to anxiety and stress.

A rapid or frequent change of phases due to changing tasks and/or attention should be avoided.

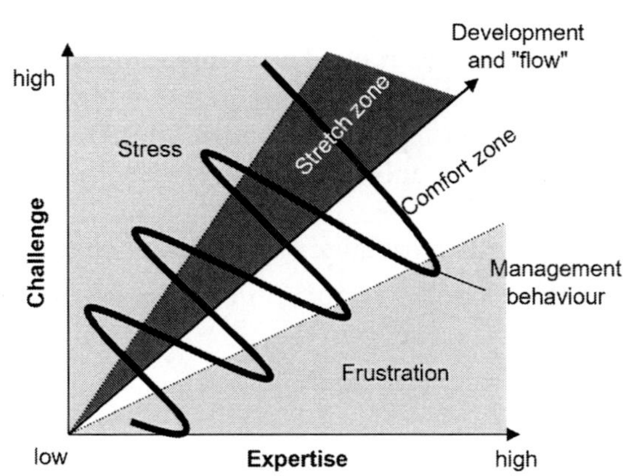

FIGURE 3.14
Managing with flow [8]. (Generated by authors.)

The trick is, therefore, to lead employees out of their comfort zone and into the stretch zone and to support them there so that a sense of achievement, new solutions, and more competent employees emerge. To this end, problem-solving tasks are assigned to team members who are less familiar with them, and they are supported by asking open and guiding questions.

As every employee is unique in terms of their skills profile and willingness to leave their comfort zone, it is important to personalise the coaching and tasks.

From a certain level of competence, employees can take on coaching tasks themselves, and their problem-solving expertise is spread more widely.

3.8 Example of Systematic Problem-Solving within DM

To illustrate systematic problem-solving in DM and coaching, an example from the CiP process learning factory is described. The production environment of CiP is first introduced, and the progression of a problem in various areas is discussed.

3.8.1 Production Environment: CIP Process Learning Factory

The Process Learning Factory is located on the campus of the Technical University of Darmstadt. Lean production training courses for companies and students have been offered and held there since 2007. A variable set-up based on real production makes adapting the environment to the respective training objective possible so that participants can directly implement their solutions. The realistic production environment is also used to conduct research in the areas of lean, Industry 4.0, digitalisation, and artificial intelligence.

The CIP "produces" pneumatic cylinders in eight standard variants as well as configurable, customised products. The cylinder bases and piston rods are machined in-house, while the other components are treated as purchased parts. The hall includes several production machines, measuring equipment, logistics systems, and an assembly area (Figure 3.15) [9].

1. Saw: cutting the raw material for the cylinder bases
2. Customised production: Production of customised heads and piston rods in sequential production
3. Lathe: Production of piston rods
4. Milling machine: Production of cylinder bases
5. Quality station: cleaning and measurements
6. Assembly and packaging: five workstations in a U-line configuration

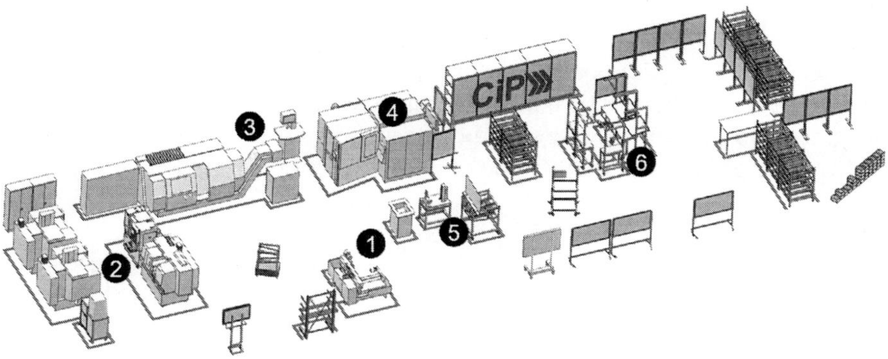

FIGURE 3.15
Layout of the CIP process learning factory. (Generated by authors.)

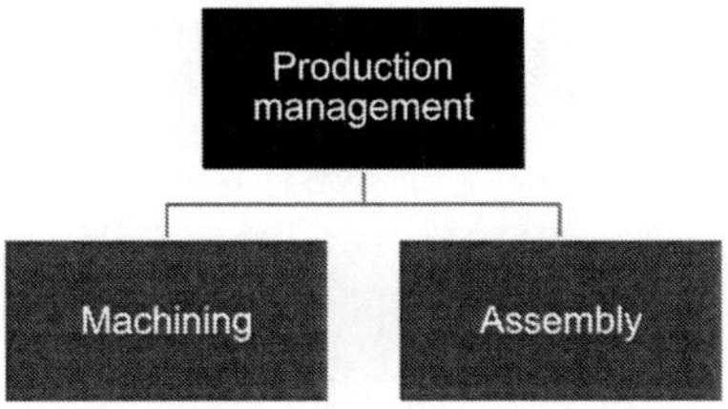

FIGURE 3.16
CIP organisation chart. (Generated by authors.)

FIGURE 3.17
Front and rear of the defective pneumatic cylinder. (Generated by authors.)

In organisational terms, the production is divided into Machining, including the measuring station, and assembly with higher-level production management (Figure 3.16).

Accordingly, there are also three DM meetings: at 8.00 a.m. in each department and at 8.30 a.m. at the production management level.

In this context, let's look at an example of problem-solving in the DM: Today, an employee discovered something unusual during assembly: The underside of a cylinder base is not as smooth as usual but has grooves (Figure 3.17).

3.8.2 Daily Meeting Assembly

The employee brings the part to the daily meeting.

The meeting starts at 8.00 a.m. Attendance, news, the previous day's performance and current measures are discussed using the daily management board. On the subject of quality, the employee mentions the defective part:

Employee
"I noticed something else today. This part doesn't have the usual smooth surface but has these grooves. I have already measured it. Instead of the expected Rz of 20 µm this part shows 85 µm – so way over the specification. The milling tool has probably broken off."

Assembly management
"Thank you for bringing that up. We don't know yet whether the milling tool is really the point of cause. Can you please describe the error in detail?" When, where exactly, and how often did you observe the error?"

Employee
"Gladly. It was today at 10 o'clock in a batch of the standard variants, and the whole underside is evenly affected. I saw it directly at my workplace in assembly. It was just this one part."

Assembly management
"Thank you. In my opinion a broken milling tool is unlikely to affect exactly one whole surface. So we do not know why this deviation occurred yet. Please let me summarize and note the deviation statement:

Roughness on cylinder base
The whole underside of one piece of the standard cylinder base exceeds the acceptable roughness (Rz) by 65 µm.
"And how should we deal with it?"

Employee
"Firstly, we should look through all the parts. It disrupts the assembly process, and we should avoid delivering a part like this."

Assembly management
"All right, so our customers won't be affected for the time being. So, we're doing a complete check. Has this happened before?"

Employee
"I had this a few weeks ago, and the others have also reported that this problem keeps popping up."

Assembly management
"Thank you, then we should solve this issue for good. I'll take it with me to the production management. Can you please give me the part?"

The team management reacted correctly to the error message. The employee ensured that a cause was not hastily concluded. Instead a complete deviation statement was provided, and the issue was prioritised. The employee was guided by questions instead of instructions, which showed appreciation.

Parallel to the discussion, the measure was noted on the board, and the deviation statement was noted on a problem card. The card is used to discuss the problem in the production management meeting.

3.8.3 Daily Meeting Production Management

30 minutes later, the production management team meets. After discussing the key figures, the assembly manager addresses the defective part.
Assembly management

> "I have brought a defective part and a deviation statement with me. This type of deviation has occurred several times before. What can we do to prevent it from happening again?"
>
> Production management
> "A part like this must never leave the production line. Do we know how something like this can happen?"
>
> Work preparation
> "The back is not machined. The tool must have broken off."
>
> Machining team lead
> "No, I don't think so. It must be a programme error."
>
> Quality
> "Then there would have been more parts. Surely the Z-axis wasn't set correctly."
>
> Work preparation
> "I stick to my guns: tool breakage is the most likely cause."
>
> Production management
> "We're not getting anywhere like this. Please take the issue with you and find a systematic solution to the problem with your team."

Firstly, the positive error culture should be emphasised in this exchange. The team is already looking for a cause without making personal accusations or introducing a quality level that only sorts out defective parts. However, the discussion is drifting away from objectivity as different opinions on possible causes are exchanged. The moderator intervenes quickly and starts a systematic problem-solving process.

The production team leader then makes an appointment with the milling staff on the shift where the error occurred to find the cause on the machine based on the facts.

3.8.4 On-Site Problem-Solving

The production manager meets with the milling cutter on-site later in the day.

Machining team lead
"Good morning, thanks for your time, we haven't got any further on the deviation "Roughness on cylinder base". I've summarised, what we know so far, in a problem specification:

Dimension		Deviation (IS)	Deviation (IS NOT)
What?	At which object is the deviation to be observed?	Cylinder base of the standard variant	Not discovered on purchase or turning parts
	What is the deviation precisely?	Rz at 85 µm not at 20 µm	
Where?	Where do we observe the deviation (geographically)	During assembly	
	Where is the deviation at the object?	On the whole underside	On the top side or on parts of the underside
When?	When was the deviation observed first?	Unknown, at least several month ago	New phenomenon
	When was the deviation observed since then? Is there a pattern?	Every couple of weeks, no clear pattern	
Extent	How many objects are affected?	Only one part every time	Several parts each time
	How big is each deviation?	Rz 65 µm over specification	

We already exchanged some ideas and suspected the milling machine (Figure 3.18). Let us challenge these ideas and see which one complies with all the dimensions of the problem.

Milling machine operator
"All of these ideas don't match with the problem specification: A program error would always affect more parts, an offset in the axis

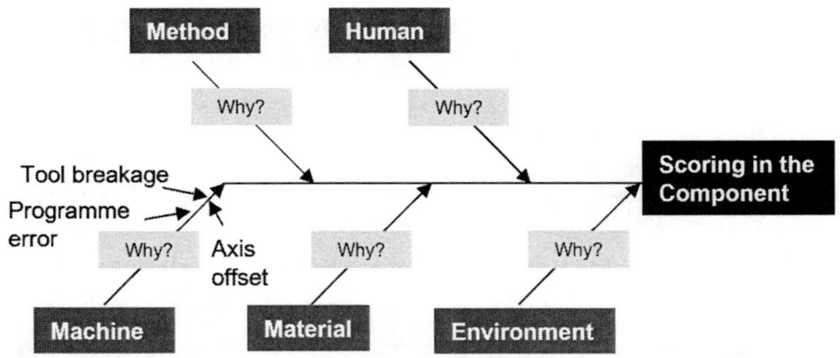

FIGURE 3.18
Ishikawa diagram for root cause analysis of the scorings. (Generated by authors.)

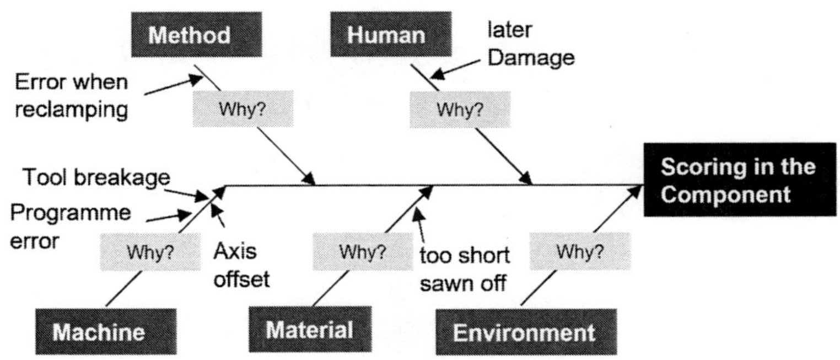

FIGURE 3.19
Ishikawa diagram for root cause analysis (extended). (Generated by authors.)

would also be a problem on the top side and a tool would brake while working on the surface, leaving only parts of it with this deviation."

Machining team lead
"Ok, then let's collect some more ideas (Result in Figure 3.19). Do they explain the problem?

Milling machine operator
"Later damage can be ruled out, that could also happen to turning and purchase parts. The reclamping process is predefined by a fixture, if it fails it would again happen to more parts at once. I think it's most likely that the material was sawn off too short."

Machining team lead
"We can test that directly - in fact, the part is 0,1 mm too thin."

Milling machine operator
"Then the milling head no longer reaches the material, and the surface from the sawing process remains."

Why was a piece that was too short passed on?

It must be the first cut. The end piece is wider for the clamping and the feed is automatic.

Why was the first cut passed on?

It's usually noticeable because it's much shorter and we sort it out straight away.

Why was this piece similar in length?

We always insert the bar from behind, where we have made a small mark.

Why is the mark not sufficient to ensure process reliability?

The bars are only very roughly toleranced as raw material, the stop at the back is not sufficient to ensure that it fits at the front.

FIGURE 3.20
Result of the 5xWhy analysis. (Generated by authors.)

Machining team lead
"I'll get the employee from the saw to join us, and we'll see how this could've happened. Good morning. We believe that the saw machine is the point of cause for this roughness on the cylinder base as parts get sawn off too short and passed on every couple of weeks. Is that possible?"

Saw operator
"We clamp the raw material from the back and then feed it automatically. After that we have to sort out the first cut by hand. Normally it is easy to spot, as it is very thin, but I remember cases, where I could not find it."

Machining team lead
"InterestingInteresting, let's get to the bottom of this via a 5xWhy." (result in Figure 3.20)

Machining team lead
"This gives us our root cause: the method of aligning the material from the back is not reliable, as the bars have different lengths. How can we avoid this in future?"

Saw operator
"We could introduce a 100 % manual measurement of the parts or lower the tolerances of our material."

Milling machine operator
"Perhaps automatic process control of the finished parts by weight would be easier."

Saw operator
"The easiest thing would be to move the stop forward, then it can't really happen anymore."

Machining team lead
"For cost reasons, I would first opt for the stop at the front of the machine, and then we'll see if that's enough."

The team jointly identified and discussed various causes. The most likely was selected and traced back to the root cause. The team leader also asked skilful questions to identify a root cause. At any point in the 5xWhy analysis, the manager could have decided that the employees should be better instructed or "take better care." Instead, a new, better process was developed.

The PLAN phase is now complete. The measures can now be distributed to maintenance and the team leaders in the meeting cascade. For a sustainable solution, not only the DO measure but also the CHECK and ACT are followed up:

DO	Design front stop and test applicability with the team.
CHECK	Testing of the process reliability of the front stop.
	Observe whether the problem occurs again.
ACT	Check whether there are similar processes in the company.
	Include in process design guidelines.

The measures will be followed up in the next few weeks on the action plans of the teams involved until it is ensured that such a deviation is not repeated on this machine or other machines in the company.

3.9 Summary: Characteristics of Successful DM

For DM to improve operational processes by developing and channelling the problem-solving skills of individuals and the organisation, many methods and people at different hierarchical levels must work together. This chapter

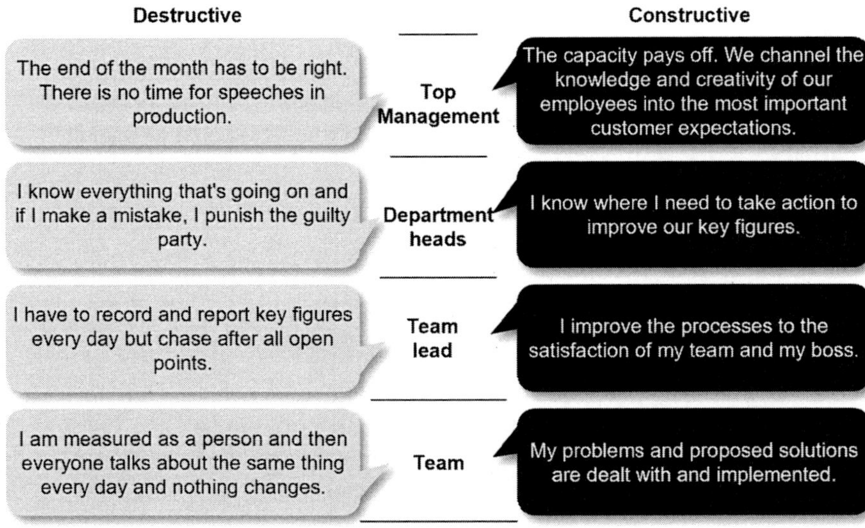

FIGURE 3.21
Destructive and constructive attitude towards DM. (Generated by authors.)

summarises the characteristics of a successful DM based on the attitude of various functions in the company.

The team, the team lead, department heads, and top management have different roles in achieving the operational goals. If all hierarchical levels recognise and use DM as a tool, it will bring the desired success – but this is not always the case. If DM is not introduced and practised correctly, a destructive attitude can prevail at different hierarchical levels (Figure 3.21).

If key figures are primarily used for monitoring purposes and improvements are not implemented, the teams reject the DM as a control tool. In this case, the team leads also see the DM as a primarily additional work. An incorrect understanding of errors in the department heads and top management often triggers this. However, if problems are solved and improvements are implemented that positively impact the KPIs, the teams and the management prosper. The department heads support the improvement activities, and the management recognises the worthwhile use of capacity.

DM is, therefore, particularly successful when measures are implemented. This not only has the greatest influence on the operational objectives, but employees also see this as the greatest motivation because commitment in the team is not achieved mainly through financial incentives or "coercion" – the prompt and binding implementation of measures has the greatest influence on participation in the CIP [10].

dDM was to be introduced at a machine parts manufacturer. The works council exercised its right to have a say and was initially sceptical about the project. The question was, "What's in it for the employees?" It turned out that the daily meetings only started at the foreman level and that team members were approached individually in advance. This meant that feedback on problems or suggestions did not reach the employees. As a result, the employees requested that the dDM system allow those involved to view the processing status of measures. This feature led to employees' support for the dDM and greater satisfaction with deviation management.

With these correlations, the condition of a DM system can be quickly analysed based on conversations with individuals. A template for a detailed analysis of a DM is presented in Section 4.6.

3.10 Weaknesses of Analogue Daily Management

If DM is such a driver of success, why is it not being used everywhere, and why should it still be digitalised? Analogue DM also has a number of weaknesses that repeatedly lead to the spread of bad attitudes towards DM.

The weaknesses of the analogue DM are the time-consuming data collection and distribution, the associated inconsistent data storage in different systems and/or levels, incomplete documentation of decisions and measures, the person-dependent understanding of methods and the lack of historicisation of key figures and deviation data.

Complex Data Collection and Distribution

Creating the necessary transparency requires a great deal of effort. Even recording and distributing daily key figures is time-consuming, and recording them at shift or hourly is hardly feasible. The effort is particularly on the team leaders, who refuse to take on too much work.

Especially if the necessary data is already maintained in IT systems, explaining the added value of additional manual maintenance to leaders is difficult.

Inconsistent Data in Different Systems and/or Levels

A parallel "real" system to the data in the IT systems is often set up on the whiteboards: "The system has reported five deviations, but they are incorrectly assigned to us. It was actually only one". Or different truths are stated for different levels: "I show the team five errors so that they take action, but I only show one to my manager so that I look better". These whiteboard options counter the principle of transparency for prioritising measures and problems.

Incomplete Documentation of Decisions and Measures
The results of the meetings are not documented in a centralised and accessible manner, but on paper based action plans. Measures are noted individually or problem cards are exchanged. Due to lot transparency over all measures, it remains easy. to delay measures and problems – and in the long term, this leads to a rejection of DM at all levels.

Person-Dependent Understanding of Methods
The success of the DM depends on moderation. The understanding and acceptance of the leadership style are usually very different. The larger the organisation, the more difficult it is to build up an understanding of leadership from a central role and to see where there is a need for further training. The DM is, therefore, perceived and practised very differently in different teams and hierarchical levels within the same organisation.

No Historicisation of Key Figures and Deviation Data
The boards are usually deleted at the end of the week or, at the latest, at the end of the month – the history of the key figures, as well as the measures and problems, is lost in the process. There is no way to learn from the past systematically.

Digital DM intends to address these weaknesses, but it has its own weaknesses, opportunities, and risks. The effects of different dDM variants are explained in Chapter 4.

References

[1] Peters, R.: *Daily management. Leading at the point of value creation*. Stuttgart: LOG_X 2009.

[2] Hertle, C.; Tisch, M.; Metternich, J. et al.: Das Darmstädter Daily management-Modell. *ZWF Journal for Economic Factory Operation* 112 (2017) 3, pp. 118–121.

[3] Leyendecker, B.; Pötters, P.: *Daily management. Leadership at the point of action*. Munich: Carl Hanser Verlag 2018.

[4] Liker, J. K.; Meier, D.: *Praxisbuch, der Toyota Weg für jedes Unternehmen. The companion book to the classic*. Munich: FinanzBuch-Verlag 2016.

[5] Pareto, V.: *Cours d'Économie Politique. Nouvelle édition par G.-H. Bousquet et G. Busino*. Geneva: Librairie Droz 1964.

[6] Kepner, C. H.; Tregoe, B. B.: *The new rational manager*. Princeton: Princeton Research Press 1981.

[7] Rother, M.: *Die Kata des Weltmarktführers*. Frankfurt: Campus-Verl. 2013.

[8] Csikszentmihalyi, M.: *Flow. The psychology of happiness*. [S.l.]: RIDER 2022.

[9] Metternich, J.: The Centre for Industrial Productivity. Your partner for lean and digital production. Internet: www.ptw.tu-darmstadt.de/prozesslernfabrik/. Accessed on 25.01.2023

[10] Landmann, N.; Schat, H.-D. (eds.): *Ideen erfolgreich managen. New perspectives, current industry examples, scientific foundations and findings*. Wiesbaden: Springer Fachmedien Wiesbaden 2019.

4

Digital Daily Management

ABSTRACT Numerous people, methods, and tools are involved in DM, and only the right interaction can lead to sustainable improvements. Digitalisation can simplify many of the method's steps, but it also changes the entire system. Therefore, critical voices often warn against the digitalisation of DM. This chapter provides an overview of the strengths, weaknesses, opportunities, and risks of dDM and how different variants of dDM address these differently. Finally, an example of a day's work in a dDM ist given.

4.1 Strengths, Weaknesses, Opportunities, and Risks of Digital Daily Management

In the context of this book, digitalisation is understood as an enabler. It forms the basis for generating, analysing, and using data for new or improved processes, products, and business models. Whether this is good or bad for DM depends on how the data and its analysis are used. This section shows the effects of digitalisation on the daily management control loop and summarises them in a strengths, weaknesses, opportunities, and threats (SWOT) analysis.

Digitalisation can significantly reduce the biggest disadvantage of analogue DM – the time-consuming manual collection of data for target/actual comparisons. This is especially true if data is already available in digital systems and must be printed and displayed for the daily meeting [1]. The target/actual comparison is often not only easier, but also better [2].

DOI: 10.1201/9781003612681-4

Users of dDM emphasise not only the time saved in the preparation of key figures but also their improvement: up-to-date, unadorned figures that can be tracked from a high aggregation level down to machine level direct attention and decision-making better. The "discussions at a much higher level" improve problem identification and solving.

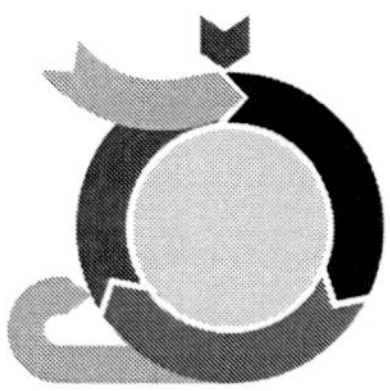

Digitalisation also has a major impact on the moderation of daily meetings. The system can specify the meeting duration and content for moderation, as can the creation of deviations from "red" key figures. A digital system can create and visualise calculated key figures more easily and can be used by managers away from the shop floor. However, this also allows managers to act as supervisors from a distance, which can lead to alienation or rejection by employees.

This risk is seen in particular by representatives of a conservative understanding of lean. They argue that a digital system runs counter to the basic understanding of lean, which emphasises local leadership, respect, teamwork, and the development of excellent employees rather than technical systems.

During a research trip to Japan, seven companies, including Toyota, were asked whether they saw a threat to the Gemba mentality through the digitalisation of production and problem-solving processes. None of the interviewees understood the question and never asked themselves it. A data basis makes it easier to take a closer look. All of the companies surveyed don't want to miss out on this opportunity to enhance their view on Gemba.

In contrast to a whiteboard, software has limited adaptability: the moderator cannot simply customise it to all needs. However, this characteristic is often seen as positive from the central perspective of a standardised procedure, as the content and methods of the DM can be better standardised and reviewed.

The data basis enables easier exploration and better interactivity so that analyses can be carried out.

In digital daily management, tracking the PDCA across all levels is significantly simplified. In dDM, the knowledge acquired does not simply remain on paper and in the heads of the people involved. Still, it is documented in a searchable and reusable knowledge repository. This is particularly important in coping with the loss of knowledge and experience that threatens to result from the retirement of the baby boomers.

For systematic problem-solving, assistants train the understanding of methods, provide guidance on procedures (e.g., problem-solving), and ensure that problems are dealt with completely (e.g., through the obligation to create and implement "Act" measures) [3].

The combination of data availability, problem-solving assistance, action distribution, and tracking paves the way for faster and more sustainable problem-solving. Toyota explicitly utilises these options (Figure 4.1).

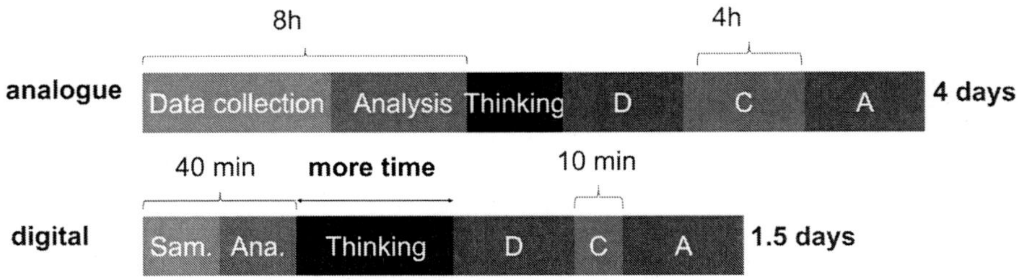

FIGURE 4.1
Acceleration of the PDCA through data availability (as seen at Toyota). (Generated by authors.)

In analogue DM, data is either recorded manually (e.g., by a process observer) or collected from various IT systems and exported (e.g., from an ERP system). Only then can employees' time be used for creative problem-solving. If measures are implemented, data must again be recorded or compiled manually to check their effectiveness before the new solution can be standardised. With digital data availability, significant time can be saved in data collection and analysis and ineffective testing, which can be used elsewhere to find solutions.

In addition, two characteristics of the digitalisation process must be considered: The costs of implementation and maintenance and the impetus provided by the implementation project to (re)energise the DM. If the digital tool also improves methodological expertise and problem-solving skills, the benefits of dDM go far beyond saving the effort of data preparation.

In summary, the following strengths, weaknesses, opportunities, and risks arise in the digitalisation of DM (Figure 4.2).

4.2 Consequences of the SWOT Analysis for Digital Daily Management

What conclusions can be drawn from the SWOT analysis? Whether the strengths of dDM are utilised and weaknesses avoided, or whether the opportunities outweigh the risks depends on the individual implementation. System design and implementation determine which aspects are realised. This chapter explains how to deal with the results of the SWOT analysis.

The system selection and implementation procedure should anchor the strengths in the digital system and utilise new opportunities. The aim is to avoid foreseeable weaknesses and minimise risks as much as possible. A SWOT analysis results in four strategies, and the conscious or unconscious

Strengths	Weaknesses
▪ Less effort required to prepare, visualise and disseminate performance data ▪ Simplified escalation and traceability of deviations and measures across all levels	▪ Costs and implementation effort ▪ Limited adaptability

Opportunities	Risks
▪ Better decision-making ability through improved transparency ▪ Methodical support through dDM system ▪ Automated notes on deviations ▪ Faster and more sustainable problem solving ▪ Development of a usable knowledge database ▪ Decision support through data analysis ▪ Re-design of the DM	▪ Lack of identification of team members with the dDM ▪ Manager does not provide on-site support but acts as a supervisor and optimiser ▪ Tracking of too many or too complicated key figures ▪ Employees cannot handle the system

FIGURE 4.2
Strengths, weaknesses, opportunities, and risks in dDM (based on [4], [5], and [6]). (Generated by authors.)

pursuit of these strategies leads to four different variants of dDM systems, all of which can be observed in practice.

Reducing Weaknesses to Avoid Risks
Risks and weaknesses should be avoided. Either the analogue DM is retained or the simplest, free digital whiteboard tools are used. However, the execution of DM does not improve.

Use of Strengths to Defend against Risks
The risks of introducing dDM arise primarily from an incorrect error culture. If only digital action management is introduced, the risks of an incorrect understanding of key figures would be avoided.

Reducing Weaknesses by Utilising Opportunities
The main weakness is the cost, but this can be recouped by saving working time and improving processes through better problem-solving.

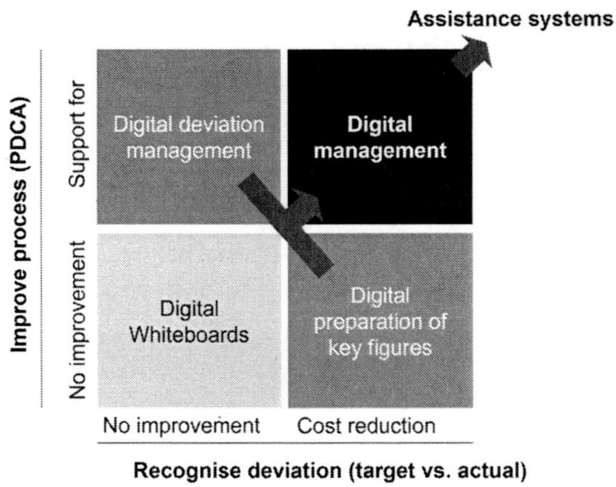

FIGURE 4.3
Variants of dDM systems in practice. (Generated by authors.)

Using Strengths to Capitalise on Opportunities
If KPI and variance management are combined, DM can be methodically supported to the greatest extent, and a complete database for assistance functions will be created.

This reveals two sub-areas of DM that can be digitalised individually: simplified deviation detection using key figures and the structuring and tracking of the PDCA. Depending on which area is supported by software, there are four typical forms of digital daily management (Figure 4.3).

4.3 Variants of Digital Daily Management Systems

The four variants shown are presented below.

4.3.1 Variant 1: Digital Whiteboards

Digital whiteboards are interactive software tools that display and edit information, drawings, and texts similarly to a conventional whiteboard. Digital pens, fingers, or other input devices can be used to write notes, make drawings, type text, and embed images and videos.

As a digital image of a whiteboard with better collaboration options, it makes sense to replace the classic whiteboards with these (Figure 4.4).

The emergence of this type of dDM is often initiated directly by team leaders, as only fairly simple software, which is usually already available,

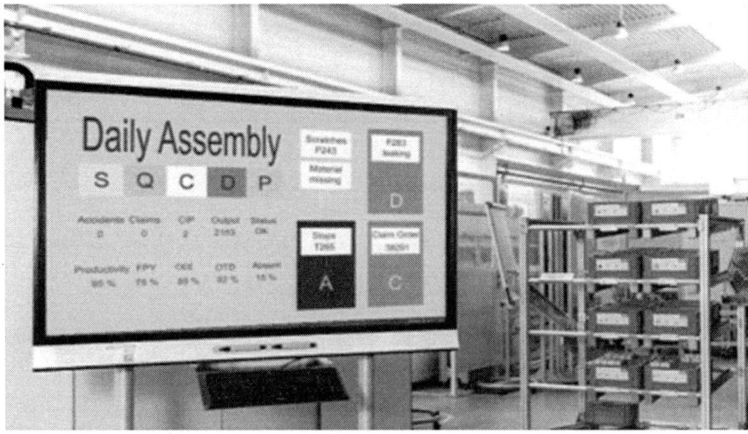

FIGURE 4.4
dDM with a digital whiteboard. (Generated by authors.)

and no external support (e.g., from IT) is required for implementation. In particular, the increase in home office regulations justified allowing as many people as possible to participate in DM remotely. Managers could help with digital whiteboards without incurring costs or major project expenses.

With the collaboration capability, tasks can be viewed from anywhere and thus be better distributed. However, the whiteboard tools have no backend and no internal logic. This means that data sources cannot be integrated, historicised data cannot be created for further use, and business logic or workflows cannot be supported.

Although digital whiteboards avoid the weaknesses of costs and lack of adaptability, they also obstruct opportunities for improving the DM, as no structured data is created or methodological support is possible. In addition, hardly any effort is saved in preparing the figures, as the figures still have to be copied daily. Due to the simplified absence of managers and support areas, a digital whiteboard can still lead to the alienation of employees or even rejection of the dDM (Figure 4.5).

> Although dDM using digital whiteboards is cost-effective and requires little effort, it hardly improves DM. It mainly enables remote participation, which is not in the spirit of the method.

4.3.2 Variant 2: Digital Deviation Management

Digital deviation management for production is similar to a ticket system. It includes the option of recording deviations (e.g., errors, faults, organisational

Strengths

- Less effort required to prepare, visualise and disseminate performance data
- **Simplified escalation** and traceability **of deviations and measures across all levels**

Weaknesses

- Costs and implementation effort
- Limited adaptability

Opportunities

- Better decision-making ability through improved transparency
- Methodical support through dDM system
- Faster and more sustainable problem solving
- Development of a usable knowledge database
- Automated indications of deviations
- Decision support through data analysis
- Re-design of the DM

Risks

- **Lack of identification of workers with the DM**
- **Manager does not provide on-site support but acts as a supervisor and optimiser**
- Tracking of too many or too complicated key figures
- Employees cannot handle the system

FIGURE 4.5
Realised properties of digital whiteboards as a dDM tool. (Generated by authors.)

problems) and then forward and track them within the organisation (Figure 4.6).

For this type of dDM, systems from other areas (especially IT) or from the Microsoft Office range of functions are often used. Digital management of measures and problems increases the sustainability of problem-solving, ensuring a high level of commitment from the workforce.

However, these tools are not designed for group meetings or performance reviews. They lack visualisation options for key figures, which can only be added with significant effort. As a result, key figures are missing as an essential prioritisation feature, both for monitoring the effectiveness of measures and for further analysis functions.

In addition, extending the software's use to other users generally incurs further costs and requires training (Figure 4.7).

FIGURE 4.6
Ticket system as dDM (schematic). (Generated by authors.)

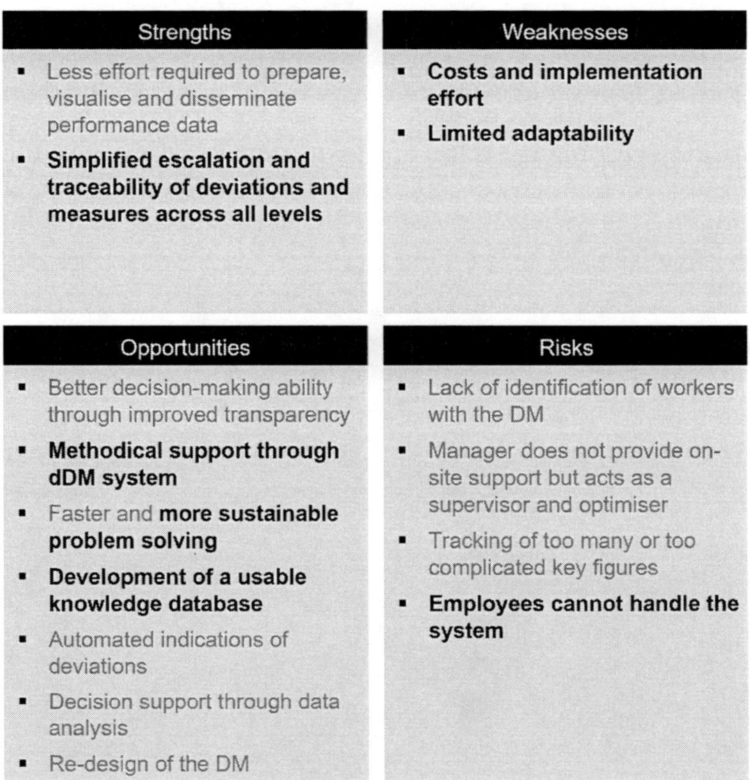

FIGURE 4.7
Realised properties of digital deviation management. (Generated by authors.)

Through digital deviation management, dDM supports the implementation of process improvements. However, the lack of integration options for key figures means that the benefits of effort savings, better transparency, automation, and data analyses for decision-making are lost.

4.3.3 Variant 3: Key Figure Dashboards

Digitalising the visualisation of key figures has immediate, measurable benefits as it takes this task off moderators. Printing out and hanging up figures already managed in digital systems is often no longer something the younger generation of managers can understand. Depending on the dashboarding tool, of which there are many, the interactivity with the figures also increases: Past data can be viewed, real-time data can be used, or detailed information can be accessed (Figure 4.8).

The ability to make decisions is increased with less effort. However, this advantage is offset by the risks associated with an incorrect understanding of errors in connection with increased transparency: Many complicated and highly aggregated key figures can be easily recorded and monitored from anywhere, quickly leading to a feeling of surveillance among employees. The purpose of KPIs – to identify and prioritise deviations and work together on solutions – is quickly lost if deviation management is not linked to the KPIs. Whether measures are taken and how the solution to the problem affects the KPIs is not recognisable or methodically promoted (Figure 4.9).

FIGURE 4.8
dDM with key figure dashboard (schematic). (Generated by authors.)

Strengths

- **Less effort required to prepare, visualise and disseminate performance data**
- Simplified escalation and traceability of deviations and measures across all levels

Weaknesses

- **Costs and implementation effort**
- **Limited adaptability**

Opportunities

- **Better decision-making ability through improved transparency**
- Methodical support through dDM system
- Faster and more sustainable problem solving
- Development of a usable knowledge database
- **Automated indications of deviations**
- **Decision support through data analysis**
- Re-design of the DM

Risks

- **Lack of identification of workers with the DM**
- **Manager does not provide on-site support but acts as a supervisor and optimiser**
- **Tracking of too many or too complicated key figures**
- **Employees cannot handle the system**

FIGURE 4.9
Realised properties of key figure dashboards. (Generated by authors.)

Key figure dashboards do not promote problem-solving itself. If the organisation is not already methodologically well developed, they are more likely to promote a false understanding of errors and reporting behaviour. Therefore, the reduced effort should be enjoyed with caution, and the organisation's maturity in problem-solving should be checked beforehand (see section 5.2).

4.3.4 Variant 4: Digital Daily Management

DDM software, which is based on a methodical understanding of DM, helps recognise deviations with little effort and anchor improvements systematically (Figure 4.10).

FIGURE 4.10
Digital DM in use. (Generated by authors.)

It includes key figure dashboards for conducting meetings, allows deviations to be recorded, and combines their processing with integrated deviation management. With supported systematic problem-solving, it ensures the sustainable elimination of deviation causes and the creation of a knowledge database. Not only are all sections of the DM control loop supported, but also the transitions between them.

The providers' expertise can also be used to improve the DM culture and develop assistance functions. This expands the organisation's problem-solving capability. This broad functionality leads to higher costs and all the risks of dDM (Figure 4.11).

> Digital daily management software supports organisational development due to their integrated workflows, which the other variants cannot achieve. That makes a real difference in improving quality, delivery times, and cost.

The higher costs of the software itself are justified by the improvement in problem-solving capability (template for amortisation calculation in Section 5.3), and the risks must be avoided through the implementation process.

Strengths	Weaknesses
▪ Less effort required to prepare, visualise and disseminate performance data ▪ Simplified escalation and traceability of deviations and measures across all levels	▪ Costs and implementation effort ▪ Limited adaptability
Opportunities	**Risks**
▪ Better decision-making ability through improved transparency ▪ Methodical support through dDM system ▪ Faster and more sustainable problem solving ▪ Development of a usable knowledge database ▪ Automated indications of deviations ▪ Decision support through data analysis ▪ Re-design of the DM	▪ Lack of identification of workers with the DM ▪ Manager does not provide on-site support but acts as a supervisor and optimiser ▪ Tracking of too many or too complicated key figures ▪ Employees cannot handle the system

FIGURE 4.11
Realised properties of dDM software. (Generated by authors.)

4.4 Working in Digital Daily Management

This chapter summarises how everyday working life is organised with digital daily management software.

4.4.1 Preparation of the Digital Daily Meeting

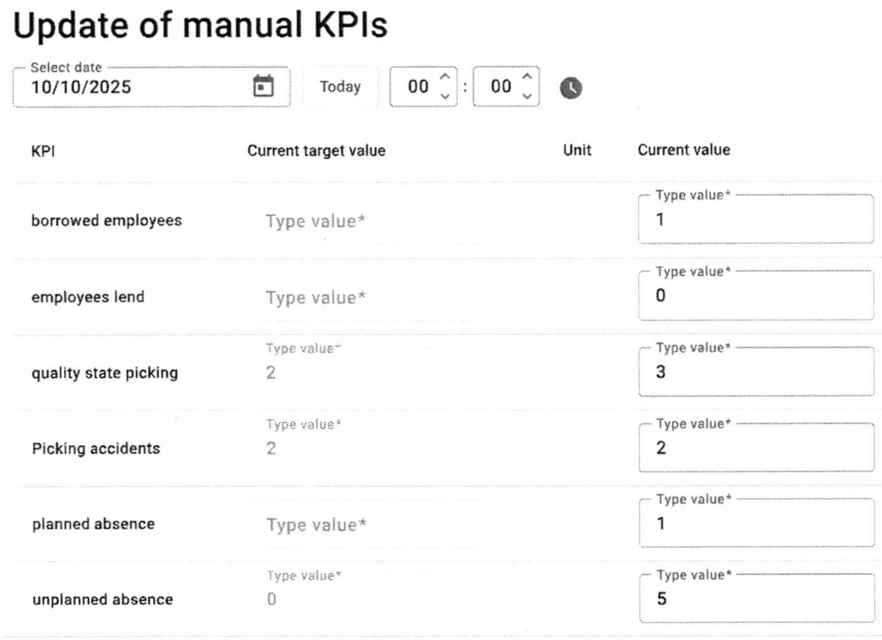

FIGURE 4.12
Central key figure update [Digital Teamboard by SFM Systems]. (Created by the authors' own company.)

In preparation for the meeting, the actual process or deviations from it are recorded. Collating key figures from various sources is unnecessary – they are already automatically transferred. Key figures for which no source system is maintained centrally by those responsible can be used directly for all levels and calculations (Figure 4.12).

The time saved is used for a Gemba Walk or a short audit. Using a short checklist, individual specifications are checked, and deviations are assigned directly to the various teams with photos (Figure 4.13).

In addition, employees can report deviations directly in the system at any time.

4.4.2 Carrying Out the Digital Daily Meeting

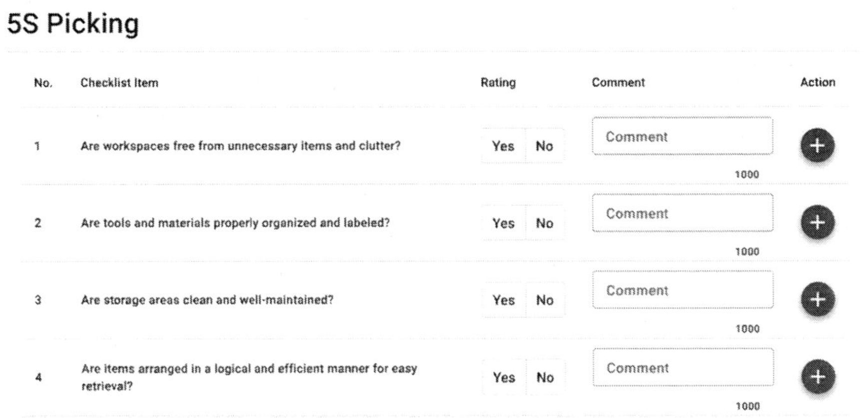

5S Picking

No.	Checklist Item	Rating	Comment	Action
1	Are workspaces free from unnecessary items and clutter?	Yes No	Comment 1000	+
2	Are tools and materials properly organized and labeled?	Yes No	Comment 1000	+
3	Are storage areas clean and well-maintained?	Yes No	Comment 1000	+
4	Are items arranged in a logical and efficient manner for easy retrieval?	Yes No	Comment 1000	+

FIGURE 4.13
Performing audits [Digital Teamboard by SFM Systems]. (Created by the authors' own company.)

With this preparation, the meeting cascade can begin. The team comes together on time – attendance and the punctual start of the meeting are recorded and conducted using the agenda.

The key figure dashboard is designed with automatic colouring and notifications so that it is immediately clear where action needs to be taken. Simple static target and limit values, as well as long-term trends, such as major deviations from a long-term average, can be used for this purpose. On this basis, the system prompts you to describe deviations (Figure 4.14).

The deviations recognised in the key figures and the prepared deviations can be processed: Are measures being taken? If so, which ones? Can a cause be identified, or has a problem-solving process been started? (Figure 4.15).

The data helps prioritise the deviations: Which deviations are particularly frequent or particularly serious? Pareto analyses are the standard for this, and further data analyses can be used to make recommendations (see Chapter 7).

The outstanding measures and problems are then discussed: Could these be completed as agreed, or have difficulties arisen? Is escalation necessary? (Figure 4.16).

Escalation rules, such as the violation of deadlines, are stored in the system and are triggered automatically.

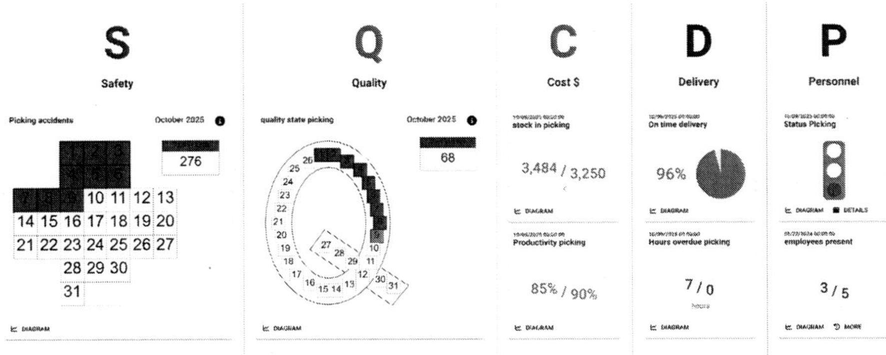

FIGURE 4.14
Key figure review in dDM [Digital Teamboard by SFM Systems]. (Created by the authors' own company.)

Deviations

Search

KPI	Description	Created/ Occurred ↓	Status	Linked Tasks
Accident in picking	Finger pinched at...	10/10/2025 18:37:06	New	
Customer Complaint	Leakage on ZP0182	10/10/2025 18:36:13	New	

FIGURE 4.15
Edit deviations [Digital Teamboard by SFM Systems]. (Created by the authors' own company.)

Tasks

✓ Current team | All teams | ✓ All | Overdue | Due today | Due this week | Due next week | Due this month

Title	Description	Due Date ↑	Assigned User	Tag	Go and See	Status
Include Marking into best practices for crane useage	Update training...	10/13/2025	Torsten Niemeyer	Training	11/10/2025	New
Add mark to crane training	Mark crane default...	11/19/2025	Torsten Niemeyer	Maintenance	11/11/2025	Started
Contain the Leak:	Place absorbent...	12/20/2025	Torsten Niemeyer	Maintenance	10/14/2025	New
Avoid safty hazards	Ensure safety by...	02/10/2026	Torsten Niemeyer	Maintenance	10/14/2025	New

FIGURE 4.16
Tracking measures [Digital Teamboard by SFM Systems]. (Created by the authors' own company.)

4.4.3 Follow-Up of the Meeting: Systematic Problem-Solving

The individuals responsible for addressing a problem gather after the meeting to solve it. The problem-solving process is supported methodically and with content. Templates are used to help describe deviations and analyse causes, while data from previously solved problems and the history of key figures are used to help in the analysis of causes (Figure 4.17).

4.4.4 Follow-Up of the Meeting: Implementation of the Measures

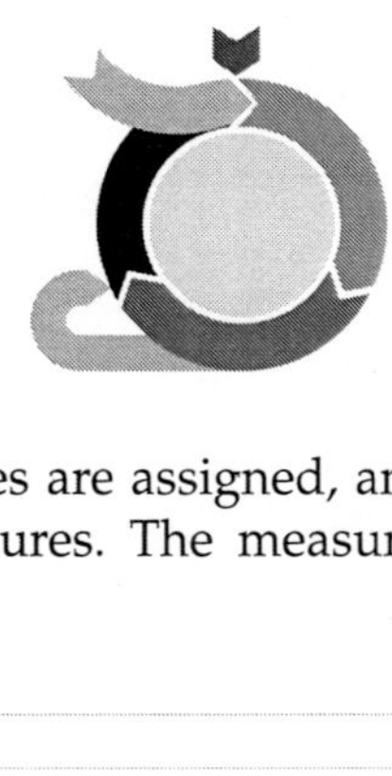

In dDM, clear responsibilities are assigned, and deadlines are agreed upon for the jointly agreed measures. The measures are not lost, even if they

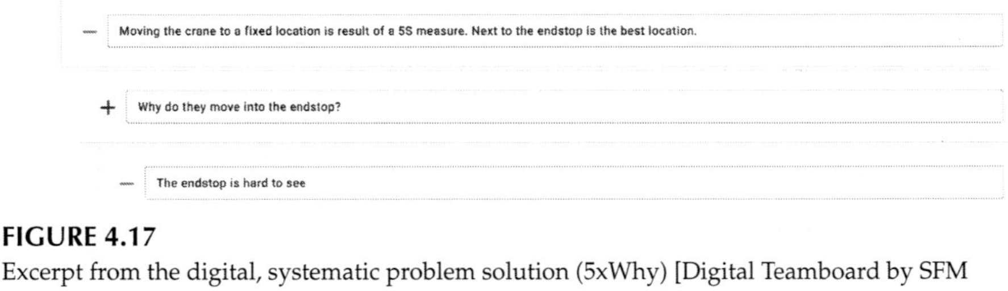

FIGURE 4.17
Excerpt from the digital, systematic problem solution (5xWhy) [Digital Teamboard by SFM Systems]. (Created by the authors' own company.)

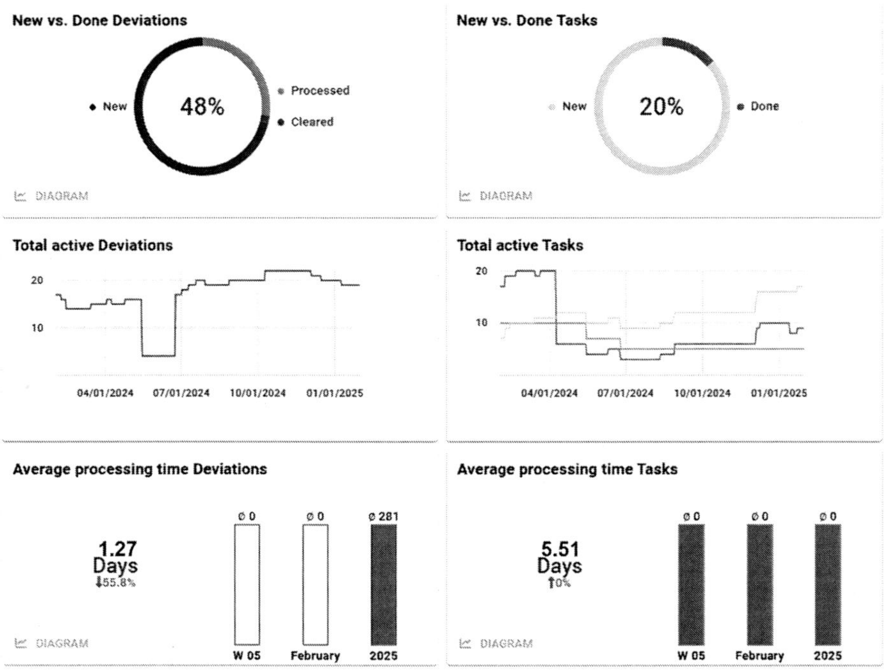

FIGURE 4.18
Analysis of deviation management [Digital Teamboard by SFM Systems]. (Created by the authors' own company.)

are escalated and distributed several times, but are always traceable. The effectiveness of action management is transparent by analysing how many deviations, actions, and problems are in which status, how large the backlog is, and what the average processing time is (Figure 4.18).

Thus, the high effort required to create transparency and distribute measures is avoided, and the system supports the correct application of the DM method.

References

[1] Iuga, M. V.: Visual communication in lean organisations. *MATEC Web of Conferences* 121 (2017), p. 2005.

[2] Lanza, G.; Hofmann, C.; Stricker, N. et al.: Auf dem Weg zum digitalen Daily management. A study on the status of real-time decision capability and Industry 4.0 maturity. 2018.

[3] Hertle, C.; Hambach, J.; Meißner, A. et al.: Digitales Daily management – Neue Impulse für die Verbesserung der Werkstatt. *Productivity Management* 22 (2017), pp. 59–61.

[4] Meissner, A.; Müller, M.; Hermann, A. et al.: Digitalisation as a catalyst for lean production: A learning factory approach for digital daily management. *Procedia Manufacturing* 23 (2018), pp. 81–86.

[5] Clausen, P.; Mathiasen, J. B.; Nielsen, J. S.: Smart manufacturing through digital Daily management boards. *Wireless Personal Communications* 115 (2020), pp. 3261–3274.

[6] Eriksson, S.; Bengtsson, L.; Samen, L. et al.: Visual management in the era of industry 4.0: Perceived advantages and disadvantages of digital boards. *International Journal of Advanced Operations Management* 15 (2023) 1, p. 1.

5

Implementation of Digital Daily Management

In many companies, DM is the most important systematic management approach in the value-adding areas. It provides the framework for a structured performance dialogue between managers and their teams. It also drives the flywheel for reactive CIP and contributes to developing employee competencies. Software that supports the DM in all phases of the DM control loop creates entirely new possibilities and can become the most important tool for operational management. Successful design and implementation, considering the strengths, weaknesses, opportunities, and risks outlined in Chapter 4, is therefore essential. The following guide shows good practices in all steps and provides templates to support the introduction of dDM software.

On the one hand, the technical systems must be planned and set up to implement dDM. However, to use the dDM system to improve the achievement of objectives within DM, the methodological expertise of employees should also be reviewed and further developed. Only with the right combination of competent employees and the right software can the organisation's problem-solving ability be improved. To achieve this, the implementation should go through five phases, each concluding with a milestone at which the progress and success of a particular phase are reviewed (Figure 5.1).

Preparation: In the preparation phase, the stakeholders are identified and prepared for the project, and the objectives are defined. The preparation phase ends with the definition of the project plan and team.

Analysis: This is followed by the analysis phase. It comprises the analysis of the existing DM and the existing IT landscape and is concluded by the creation of a specification sheet. This phase answers the question of what exactly is to be improved and which technical systems will be used as a basis. Its outcome influences the course of the most extensive phase: the design.

Design: In the design phase, methodical improvements are made to the DM processes and the technical system is created. A make-or-buy decision must be made as to which system should be procured or developed, which is then integrated into the IT landscape.

 DOI: 10.1201/9781003612681-5

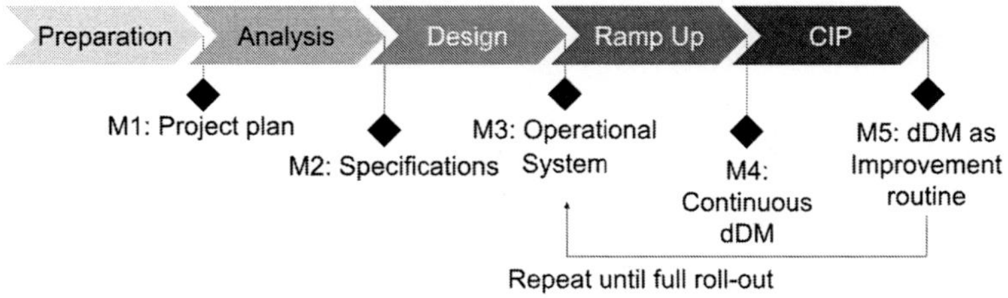

FIGURE 5.1
Five phases of dDM implementation. (Generated by authors.)

Ramp up: To ramp up the dDM software, a pilot area is first selected to test the system, and initial adjustments and improvements are implemented. The introduction requires training in the use of the software and the method, as well as the installation of the hardware and software at all meeting locations. After the go-live, the dDM can be used productively in the pilot area.

CIP: The system is continuously improved based on the experience gained from ongoing operations. The achievement of dDM targets is measured at regular intervals. It may be necessary to improve methodological expertise or adapt software features. In this way, the dDM is established as a long-term improvement routine.

DDM should initially be introduced to a pilot area in which classic DM is already the most developed. In such an area, there is a clear idea of the requirements, and the project is less likely to be delayed by prior methods and/or KPI development. Furthermore, the pilot area should have several teams from different management levels to experience the benefits of aggregation and measure management. Care should be taken to ensure that all teams in a management area participate so that the key figures of these teams can be aggregated consistently at the next higher level. If this is not consistently possible, those teams in a management area still holding their DM meetings in analogue form should manually enter their key figures in the dDM system during a transition phase. It is important to avoid hybrid solutions in which higher-level teams have to process key figures and measures on paper and digitally in parallel.

The pilot area also acts as a beacon: employees and managers from other areas participate in dDM meetings, gain an impression, and provide feedback if necessary. This ensures that the selected solution suits other areas and that sceptical stakeholders are involved.

The roll-out only occurs once the dDM has been successfully introduced in the pilot area and any necessary improvements have been completed. The experience of completed projects shows that the strengths of dDM become more apparent as teams are involved. Significant effects can be expected, particularly in the areas of key figure aggregation and action management.

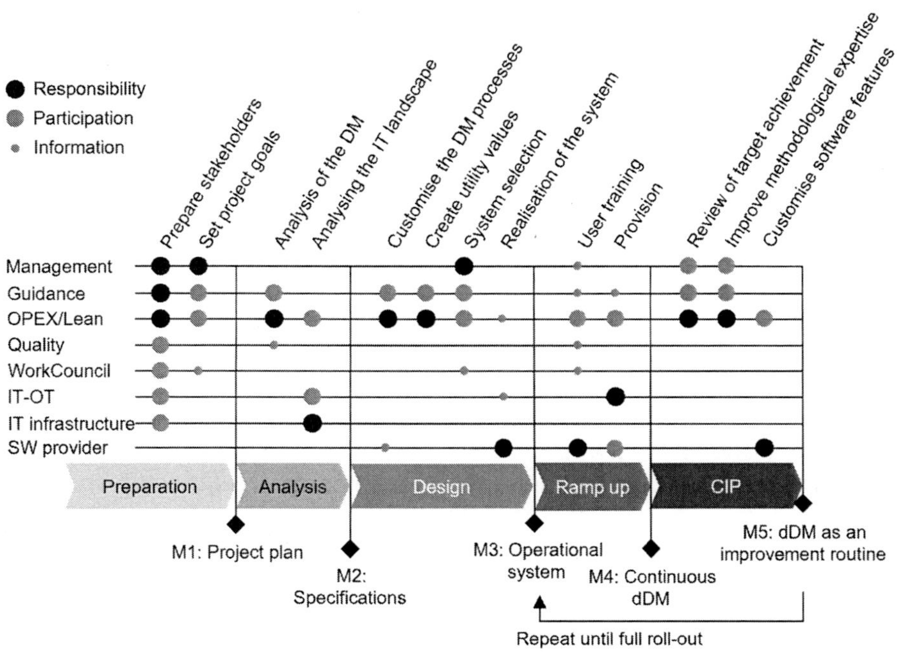

FIGURE 5.2
Responsibilities in the project phases. (Generated by authors.)

If the roll-out encounters areas where DM is not yet established or is insufficiently established, more effort is required for methodological training and coaching. However, this is also an opportunity for the entire production organisation and management, as the maturity level of DM is also aligned in other areas, and the problem-solving capability of the organisation is strengthened. Sections 5.2 and 5.3 show how the maturity level of the DM can be determined and which measures result from this. The pilot area also serves as a learning object for all other areas.

Even after a complete roll-out, dDM, like any other process, should be subject to a CIP. A routine for regular manoeuvre criticism should be established at every meeting level using key figures and checklists to continuously improve the procedure and cooperation in dDM.

Various functions are required in the phases to implement the project: Management and middle/lower management are critical to the successful introduction and operation of dDM. There needs to be a clear commitment to the goal of introducing dDM. In the best-case scenario, the digitalisation of the DM by a certain date is included in the manager's target agreements. Management must also promote the project goals and provide the funding and capacity for the required functions. The day-to-day work of the operational management level is significantly influenced by the dDM. Depending on how deeply a DM culture is already anchored, intensive coaching by internal lean managers (often OPEX team) or recognised external consultants may be required. The system must meet their needs, and the lean managers must ensure that the management culture continues to improve in the direction of sustainable problem-solving by introducing dDM. As the quality department is responsible for internal process standards and problem-solving processes, it should be involved at an early stage. It is important to record and integrate the quality requirements for dDM software so that they can be pulled together later. Operational IT, the IT infrastructure, and a software provider are required for implementation and provision. Finally, the works council should be involved, as they might have a say. The involvement of the individual functions varies in the individual phases (Figure 5.2).

The five phases are described in detail below and supplemented by good practices, decision-making aids, and templates for all project steps. A poster for the entire implementation process is available with our templates, link at the start of the book.

5.1 Preparation

For preparation, the stakeholders must first be identified and consulted. This allows the objectives and those responsible to be defined.

5.1.1 Prepare Stakeholders

The aim is to identify those stakeholders with a high level of interest and influence and involve them in further decision-making. Ideally, a committed team from several hierarchical levels can be found to make decisions on behalf of their function. Particular attention should be paid to managers and opinion leaders who are critical of the digitalisation of DM for methodological or financial reasons. Recognising, involving, and convincing these people at an early stage facilitates the project's progress [1].

Good Practice

A successful implementation project requires representatives from the following functions: management, operational management, OPEX, IT infrastructure, IT operations, works council, quality, and possibly a software provider (see Figure 5.2). Depending on the company's size, some functions may overlap; for example, operational management may also fulfil the OPEX function if there is no dedicated staff. The following checklist has proven useful in finding and collecting the right stakeholders. This checklist can be used to identify and address the necessary functions in your own organisation (issues to be clarified). It also highlights the risks if this is not done.

Stakeholder	Topics to be Clarified	Risk
Top Management	• Who is sponsoring the project/can provide the budget? • Which people need to be consulted for a management decision?	• Successful pilots will not be rolled out further • Parallel projects in different areas • No financing
Department Heads	• In which area can we pilot/implement digital DM? • Is hardware equipment (screens) available?	• Some areas are not reached via the cascade • Delaying the schedule
OPEX/ Lean	• Who will be responsible for project management (pilot/roll-out)? • Are resources available? From when?	• Method is not strengthened • Lack of capacity in the project team
Quality	• Are there QM processes to be observed?	• Parallel systems for deviation processing and problem-solving
Works Council	• Is the planned project subject to co-determination? • Do you need a company agreement?	• Project interruption by the BR • Rejection by employees
IT OT	• Are the contacts for implementation defined and informed? • In what timeframe can this be realised?	• No capacity for realisation
IT Infrastructure	• Where should the dDM software be hosted? • Do you have to fulfil IT security guidelines?	• Project interruption or stop due to IT security guidelines

If relevant stakeholders are made aware of the ongoing project too late, it can lead to delays and stops. This may be because the capacity of the right IT employees has not been reserved (or they have to be found first), operational managers refuse to accept the system, or the works council raises objections.

Furthermore, the concepts of DM and dDM often differ considerably in different hierarchical levels and areas. For decision-making, stakeholders must develop a common understanding of the underlying methods of DM (see Chapter 3) and the different design options of dDM (see Chapter 4).

In companies with up to 1,000 employees, the people relevant to the project usually know each other well. An overview of the stakeholders and their attitude to dDM can be quickly created "off the top of the heads". In Small and Medium-sized Enterprise (SMEs), a production or plant manager can often make decisions alone and does not need to coordinate with others. However, the larger the company, the greater the effort required to identify and assess the relevant stakeholders. For corporate groups, it is advisable to carry out a systematic stakeholder analysis (e.g., according to Krips [2]).

In group structures, there are many managers at the same hierarchical level, e.g., plant managers. In general, they are not dependent on each other's decisions. However, a dDM should be standardised across the group in the long term, making early coordination necessary. Identifying the relevant people at the management level and the right IT experts internally later on can take up to several weeks and should, therefore, be started at an early stage.

Four types are typically found in the various functions, each of which can be won with its own arguments in favour of the dDM:

DM Denier

This group considers even the short meetings between employees in the DM to be a waste of time. It is, therefore, first necessary to explain the advantages of transparency and an open error culture in DM (see Chapter 3).

dDM Sceptics

Sceptics of the digitalisation of DM usually argue with the risks of dDM, which are that digital tools lead to an alienation of employees and managers from what is happening on the shop floor. In their view, the digitalisation of DM leads those involved to focus more on the digitally provided information and the associated medium than on the real things happening on-site.

If classic DM is understood and practised in the organisation, this fear is unjustified, as managers use the time saved in data handling to solve problems. If DM is not yet widely established, there is a real risk that the tool will not be used correctly. In this case, however, the methodological competence in the organisation can be strengthened by the implementation project – if the right tool is introduced correctly. If the software supports the entire method, the labour savings in data handling will increase acceptance and meeting, and problem-solving assistants will make implementation easier. The introduction to the tool should be accompanied by method and moderation training (include this in the project plan). The digitally predefined processes support the newly taught processes, making adhering to them easier in everyday life.

In addition, the dDM system can be used to view how the system is being used, enabling methodical training to be planned more precisely.

Followers

This group follows the decisions made. They do not have to be included in the decision-making bodies.

dDM Supporters

This group sees the benefits of digitalising DM and would like to improve KPI dialogues, deviation management, and improvement activities. They support the implementation but may have different ideas about the project plan or software.

The result of the stakeholder analysis is an overview of the people who are relevant to a successful dDM project – supporters and hinderers.

5.1.2 Set Project Goals

The project objectives can be formulated based on a shared understanding of what digital solutions can achieve in DM: What should the system improve? This results in initial requirements for the dDM software and the system's success can be measured against these.

Good Practice

Typically, the motivation for dDM lies in improving DM processes: the aim is to create better transparency about processes and ongoing problem-solving activities with less effort. In order to justify the costs of dDM, this abstract objective can be translated into measurable variables:

- Safe time for moderators (e.g., 45 minutes per day for data preparation).
- The processing time for measures is reduced (e.g., by 30%).
- Target achievement of selected key figures increases (e.g., increase in OEE or productivity by 5–20%).

These concrete goals also enable an initial assessment of the benefits.

5.1.3 M1: Project Plan

With an overview of the stakeholders and the project objectives, the project plan can be defined with the responsible parties and time frame.

Good Practice

For reasons of simplicity or with the intention of changing familiar processes as little as possible, it is very tempting to map an existing DM in dDM software. This greatly shortens the preparation and analysis phases, and the specification and piloting of the software can be completed in just a few weeks. However, such an approach misses the opportunity to recognise and address DM's weaknesses and design the new DM control loop in such a way that the strengths of the DM software are fully exploited.

A dDM system offers significantly more options for interacting with and displaying key figures than an analogue whiteboard. Accordingly, key figures can also be defined and displayed differently. For example, the raw data of calculated key figures can be called up or summarised in various graphics. The key figures should be revised, taking these possibilities into account.

Also, the key figures are often not the only thing that should be revised when introducing a dDM system. The status of DM can vary considerably from area to area. Therefore, a broad analysis of the current status of DM should be carried out, and the implementation of dDM should be used to improve methodological knowledge.

> Many companies that had already been using analogue DM for years reported after the dDM project – that knowledge of the method was significantly worse across the board than expected.

These activities result in the following project plan for selecting and introducing an already available dDM (Figure 5.3).

In addition to the project plan, the project organisation must also be defined. In practice, a division into working level, project management, and steering committee has proven successful (Figure 5.4).

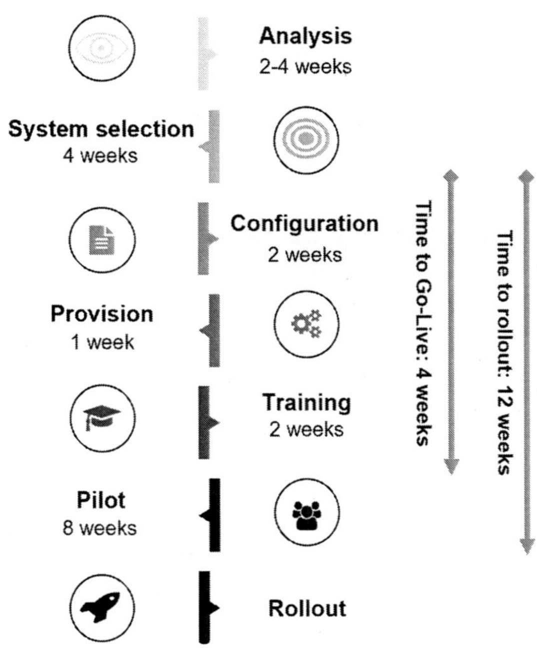

FIGURE 5.3
Example of project plan. (Generated by authors.)

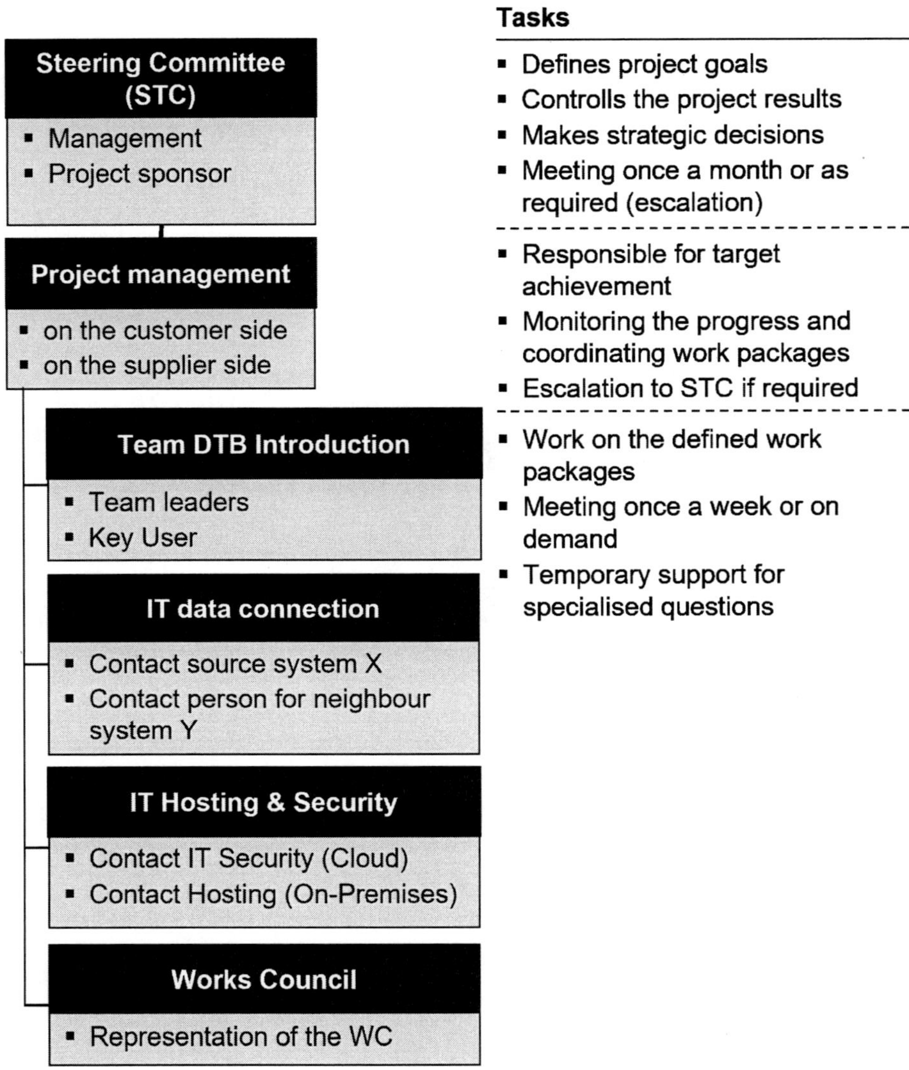

FIGURE 5.4
Project organisation for the implementation of dDM. (Generated by authors.)

5.2 Analysis

An analysis of the existing DM is recommended to determine whether the area is ready for digitalisation. And if not, it is important to understand the need to redefine current processes. From this, the requirements for the DM software are derived in detail. The aim is to tailor the system to the needs of the users. The existing IT infrastructure must also be analysed for technical reasons.

5.2.1 Analysis of the Existing Daily Management

If the teams and the various management levels have worked together in a spirit of trust in the past, there is no need to worry that the system will be used for control purposes in the dDM. Otherwise, the error culture must be developed before or with the implementation of the dDM [1].

The principles of CIP and the methods of successful DM were explained in Chapters 2 and 3. The following audit (Figure 5.5) has proven itself. It is designed to determine the status of the DM by asking a few selected questions and prioritising the necessary steps for the implementation of the dDM. It is divided into measurements of the maturity level of the shop floor culture (6

Daily meetings are organised for all levels (from teams to management).		completely fulfilled	rather fulfilled	rather not	not fulfilled
1	*Comment:*			1	
There is an agenda for the meeting.		completely fulfilled	rather fulfilled	rather not	not fulfilled
1	*Comment:*			1	
Meeting takes place according to the agenda.		completely fulfilled	rather fulfilled	rather not	not fulfilled
1	*Comment:*		1		
The defined participants are present to the meetings.		completely fulfilled	rather fulfilled	rather not	not fulfilled
1	*Comment:*	1			
Managers can explain the procedure & objective of meetings.		completely fulfilled	rather fulfilled	rather not	not fulfilled
1	*Comment:*	1			
Managers can explain the roles & responsibilities in the team.		completely fulfilled	rather fulfilled	rather not	not fulfilled
1	*Comment:*	1			

FIGURE 5.5
Extract from the DM audit area "shop floor culture". (Generated by authors.)

items), the key figure definition (7 items), deviation management (8 items), and systematic problem-solving (6 items).

The template is designed to be filled in when observing a DM meeting. Only some questions should be clarified in advance (dark grey) or can be asked of the moderator after the meeting (light grey). The assessment is carried out with a "1" under the applicable assessment from "completely fulfilled" to "not fulfilled". The complete audit is available with our templates, link at the start of the book.

The evaluation provides an overview of the status of the DM. The ratings are assigned the numerical values 0 to 3 and averaged within the categories. A category above "2" means that it is already established; a "3" would be perfect, and values below "2" indicate a need to catch up.

DM Culture Smaller Than "2"

If there is not yet a culture of regular shop floor meetings, this should be established before digitalisation. The teams are often quick to ask for digitalisation to reduce their effort. Requirements can then be better formulated and content better specified.

Key Figure Definition Less Than "2"

The key figures used are unsuitable for focusing on the most important problems. Therefore, they should not be transferred directly to a digital system; they should be revised first. During the specification of a dDM, key figures are described in detail. This overview can be used to revise the selected figures. A dDM project can, therefore, be started – but more time should be planned for the specification, which should be used to revise the key figures.

Deviation Management Smaller Than "2"

The processes for initiating and tracking measures do not function adequately. A digital DM can provide significant support here, as key figures and deviations can be linked systematically, and measures remain traceable. A low value in the area of deviation management also increases the benefits of dDM: faster processing of more actions improves processes and leads to increased productivity. A higher value can, therefore, be assumed for an amortisation calculation – if the selected system supports this area well.

Systematic Problem-Solving Smaller Than "2"

Causes are not eliminated, and problems are not solved sustainably. This value is low for most companies. On the one hand, the same applies here as for deviation management: digital assistance helps with this process. It improves production if the software offers this. However, further methodical training and often the creation of better process guidelines are also necessary. The area should also be supported after the implementation of dDM.

An audit result (Figure 5.6) shows if the DM culture and key figures need to be improved at the start and how much potential there is in the organisation for deviation management and problem-solving.

From this (typical) audit result, it can be concluded that the dDM project can begin; the key figures should be adjusted in the project plan, and a tool should be selected that supports deviation management and problem-solving. The calculation of the project should consider that the production processes will improve due to the improved deviation management (template for the amortisation calculation in Section 5.3).

5.2.2 Analysing the IT Landscape

The IT landscape must be considered for technical and organisational reasons. In organisational terms, the operational managers must clarify which data from which source systems are relevant and what the separation and cooperation with neighbouring systems should look like. From a technical perspective, it must then clarify how data will be transferred, where it will be processed and how the dDM will be operated.

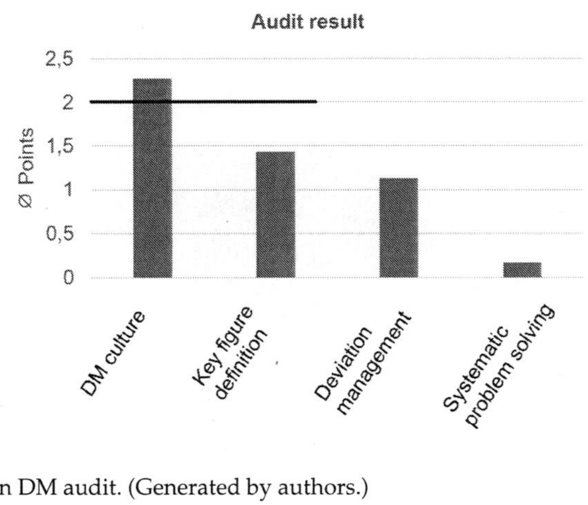

FIGURE 5.6
Typical result of an DM audit. (Generated by authors.)

Neighbouring Systems

Neighbouring systems are systems that are used in production and have links to dDM. The typical neighbouring systems to a dDM are the source systems of key figures, existing task management systems (e.g., SAP quality module, Atlassian Jira, or QUENTIC for safety issues) and specialised systems (e.g., Microsoft Power BI for reporting, systems for audits, resource planning, or qualification management) (Figure 5.7).

For these systems, it must be determined which ones will be replaced or how the cooperation or separation will be organised.

The data availability of key figures in source systems varies greatly: Existing ERP (Enterprise Resource Planning), MES, or production data acquisition systems are good databases, but often, the data is only available in MS Office or has yet to be collected. The relevant source systems must be named, and their interfaces checked (e.g., REST interfaces, access to SQL databases, Office documents, or OPC UA connectors from machines). The necessary interfaces are a selection criterion for procuring the dDM software.

In dDM, deviations are to be recognised and problems solved – this also makes action management necessary. Existing systems in which measures are also managed can either be connected or replaced. The quality system is often part of certifications and is operated by one department. Here, it is often sufficient if the dDM can display current measures. It is also advantageous if the dDM can trigger processes in other systems or import measures from other systems.

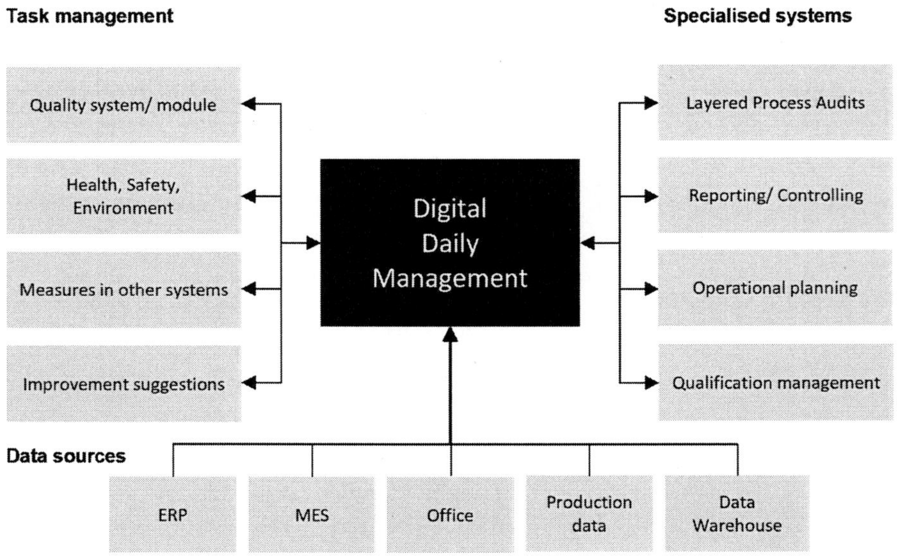

FIGURE 5.7
Typical neighbouring systems of a dDM. (Generated by authors.)

Finally, specialised systems must be considered, as they are often implemented in many companies as individual, special solutions. Depending on the company, functionalities such as audits, qualification matrices, resource planning, or reporting systems may either not be used, not be recognised as part of DM, or be completely covered by specialised software. Whether and how these functions should be included in dDM or how the interfaces should be defined should also be part of the specifications.

A template for a specification sheet can be found with our templates, link at the start of the book and is described in the Chapter "M2: Specification sheet".

Operating Mode

The dDM software can be hosted (operated) in three different ways:

Own data centre (on-premises): The company's own IT infrastructure provides a (virtual) server, and the dDM software is operated there.

Own public cloud (Own Cloud): The dDM software operates on the company's own public cloud instance, e.g., Microsoft Azure, AWS, etc.

Cloud provider (Supplier Cloud): The software provider covers the operation of the software. The dDM software is purchased "as a service".

Operation on a provider's cloud is usually the most favourable option. It must be clarified whether this is permitted for the planned data. Otherwise, alternatives such as internal clouds or servers must be provided (Figure 5.8).

Depending on the selected operating mode, the access rights must then be clarified: Where is the data of the source systems located, and how is the access of the dDM system regulated? Or, in the case of an on-premises solution, how can software providers access internal servers to carry out maintenance work?

The analysis of the IT landscape results in requirements and their prioritisation for the dDM system: Does it support the preferred interfaces and the operating mode? Does it cover the right processes? How good is the transition between dDM and specialised systems? These requirements are summarised in the specifications.

5.2.3 M2: Specifications

The analysis phase ends with the compilation of a specification sheet. It should include the necessary features for methodological support, data connection options, and IT security requirements.

A customisable template for a specification sheet is available with our templates, link at the start of the book.

On Premise	Own Cloud	Supplier Cloud
Advantages	*Advantages*	*Advantages*
• Operation in the same network as data sources and neighbouring systems facilitates data exchange • Currently the only operating option in isolated environments (e.g. pharmaceutical industry)	• Operation in the same network (on the cloud) as data sources and neighbouring systems facilitates data exchange • It is easier to implement your own IT security requirements	• Little effort required to set up and maintain the software • Operation and maintenance is handled by the software provider
Disadvantages	*Disadvantages*	*Disadvantages*
• Internal company costs for IT infrastructure and system support • Effort on the part of the software provider for updates and troubleshooting	• Internal company costs for IT infrastructure and system support • Effort on the part of the software provider for updates and troubleshooting	• A connection to the data sources must be established, e.g. via a site-2-site VPN connection
Suitable for	*Suitable for*	*Suitable for*
• Critical infrastructure that must not be accessible from outside (pharmaceuticals, defence, etc.)	• High IT security requirements for network connections	• Software-as-a-Service dDM software with short update cycles

FIGURE 5.8
Comparison of the operating modes. (Generated by authors.)

The requirements are divided and listed in detail in data connection, key figure visualisation, key figure interactions, deviation management, meeting support, individual views, process control, operation, analytics, operating mode, support, and others.

The design phase can be started with the defined requirements. The template also offers the option of evaluating five alternatives in parallel. It can, therefore, be used for system selection. For this purpose, the degrees of fulfilment (scale, e.g., 1–5) of the respective alternative are entered in the empty fields. The respective utility values are calculated in the last line (Figure 5.9).

5.3 Design

The design phase is divided into a methodological part and a technical part. Depending on the analysis of the dDM (see Section 5.2), it may be necessary to develop elements of the DM before or with the introduction of the dDM. The requirements specification forms the basis for the system selection.

Title	Description of requirements	Priority
Data connection		
System connection	The actual and target values of key figures can be imported from different systems.	
Office	Key figures can be imported from office programms.	Must
SAP	It should be possible to import key figures from SAP.	Should
REST	Key figures can be imported from REST interfaces.	Must
OPC UA	It should be possible to import key figures via OPC UA.	Must
SQL	Key figures can be imported via SQLqueries .	Must
Dynamic Target values	The target value of key figures should also be importable from data sources.	Must
Key figures visualisation		
Aggregation	The system must be able to calculate values from various datasources.	
Calculated key figures	Determine key figures from several individual values (e.g. OEE)	Must
Aggregation in hierarchy	Aggregation of values from multiple teams for higher levels by forming totals, averages or passing on of the maximum.	Must
Letter visualisations	Monthly overviews in the form of letters or a safety cross must be supported.	Must
Numerical value	Individual numerical values must be supported.	Must
Line diagram	Line diagrams must be supported.	Must
Bar chart	Bar charts must be supported.	Must
Pie chart	Pie charts must be supported.	Must
Recognising Deviations	Key figures should automatically change colour when limits are violated	Must
Report function	It should be possible to create views that contain key figures aggregated over a selectable period of time.	Should

FIGURE 5.9
Extract from the provided specifications sheet for dDM. (Generated by authors.)

5.3.1 Changes to Daily Management Processes

The analysis of the current status of the DM has revealed the areas in which the DM must or can develop further before or through digitalisation. The necessary steps need to be implemented as the next step.

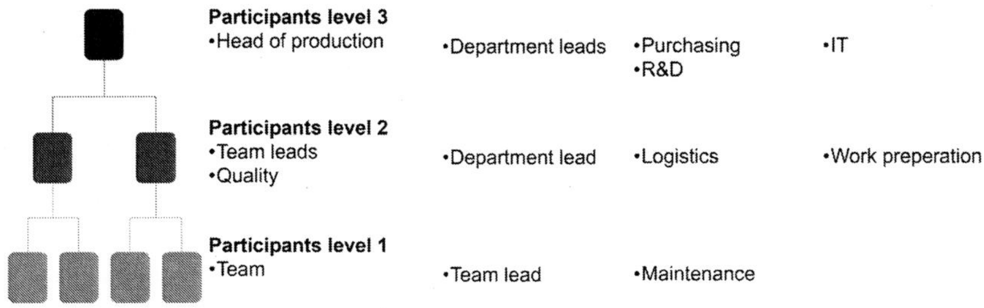

FIGURE 5.10
Result of a cascade analysis. (Generated by authors.)

Establish/Improve Meeting Culture

To redesign the meeting cascade, it is helpful to record the organisational chart of the meetings and name the respective participants (Figure 5.10).

It is often noticeable that some departments do not participate in regular communication or only at a very high level and do not participate in solving daily problems (e.g., IT or design/development). It is important to put the necessary contact people in the right places in consultation with all those involved.

Next, it is advisable to introduce rules between the levels as to when issues are escalated – especially if deadlines are exceeded (see Section 3.5).

This provides a set of rules for who attends which meetings, when, and how they work together – compliance with which can now be demanded *via* the management.

This is how regular communication is established, and appropriate key figures can now control it.

Establish/Improve Key Figure System

Chapter 2.2 has already introduced the basics of good key figures – these must now be applied to your own company.

To this end, KPIs for safety, costs, quality, delivery, and personnel are initially selected and possibly supplemented by "problem-specific" KPIs. Problem-specific KPIs measure the progression of a generally known problem to visualise its extent and to find the success of measures. Typical examples are missing parts or poor quality of submitted documents.

After defining key figures, they are distributed to the individual teams and workstations. Specific target and limit values are derived from the history or, better still, from existing process specifications. Once the aggregation rules between the levels and those responsible for recording and achieving the respective values have been defined, the revised KPI system can be gradually transferred to the communication cascade (Figure 5.11).

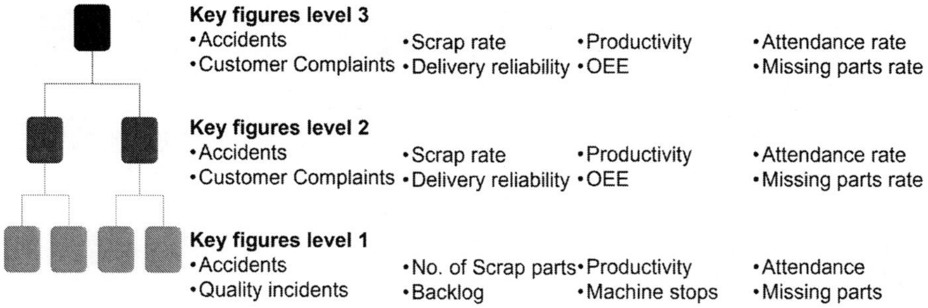

FIGURE 5.11
Overview of key figures (simplified). (Generated by authors.)

Improve Deviation Management

Poor deviation management is characterised by no reported deviations or measures not being implemented. The KPI system should show deviations; otherwise, the targets can be tightened further. The implementation of measures should be demanded *via* the meeting cascade. The digital management of measures can significantly simplify this.

Systematic Problem-Solving

The dDM system will improve systematic problem-solving if corresponding features are listed in the specifications. With the transparency and analysis options in dDM, important problems quickly become apparent. They should be dealt with systematically by the team members guided by the system (for the first problems also by mentors). Supporting specific cases is usually more successful than conducting abstract training sessions. It can be provided by external consultants, method champions (e.g., from the lean team), or managers. If internal coaches are used, it must be ensured that they have the necessary professional and interpersonal skills.

5.3.2 System Selection

The system selection can be started in parallel with the methodological reorganisation. In this step, a decision is made about what type of system should be used and whether it should be created in-house or purchased (make-or-buy). This leads to eight basic options (Figure 5.12).

The strengths, weaknesses, opportunities, and risks of the four system types have already been explained in Chapter 4 and are briefly summarised in Figure 5.12. It is often possible to create (make) digital whiteboards, action management, or key figure dashboards from existing systems (MS Office, ticket system, low-code platform) at low cost. Office programmes or systems

	Digital Whiteboard	Digital deviation management	Digital Key figure dashboard	Digital Daily Management
	– No analysable data – Enables remote meetings only	– Lack of support for meetings – Limited analysability	– No follow-up from key figures possible – Limited analysability	✓ Methodological support ✓ Detailed analyses possible
Make	✓ Existing systems can be used (e.g. MS Teams, Miro, Mural, etc.)	✓ Existing systems can be used (e.g. ticket system from IT) – Limitations in functionality	✓ Existing systems can be used (e.g. MS-Power BI) – Expertise with few employees	✓ Can be fully customised – Coordination very lengthy – Considerable development effort – several years until go-live
Buy	✓ Fast implementation	✓ Focus on improvement	✓ Can be quickly implemented as an add-on to existing systems	✓ Utilise the provider's expertise ✓ Go-live in a few weeks ✓ Continuous updates – Costs

FIGURE 5.12
Options for system selection. (Generated by authors.)

from other areas of the company can be adapted. However, the functional scope of these programmes is limited and generally cannot cover all the requirements of a digital DM. It should also be noted that the expertise for maintaining and developing the system lies with individual employees. This is risky if the employee leaves the company or even the position. In addition, the creator is usually not from IT but from a specialist department, which can affect the quality of the application.

> In-house solutions from Office programs, for example, seem inexpensive initially. However, with increasing size and complexity, they have become more and more maintenance intensive, and maintenance can often only be carried out by a few employees. System stability and key features are lacking in the long run.

In the past, large corporations, in particular, have endeavoured to create DDM systems to replicate their tried-and-tested DM on a one-to-one basis. In this way, a dDM can be customised to the company's needs – but development often takes a very long time. The ideas from the management level have to be standardised, translated into specific functions and designs, and then programmed. The conceptualisation alone takes several months, and years

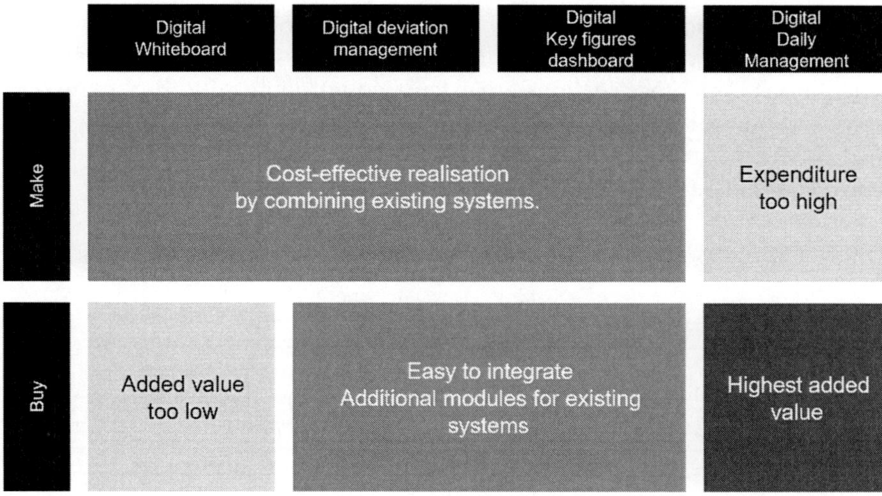

FIGURE 5.13
Frequently selected alternatives.(Generated by authors.)

can pass before a go-live. In recent years, more and more providers of dDM software have entered the market, and the products continue to mature. It is, therefore, becoming increasingly difficult to create better dDM software than what is available on the market. The development work required for this can hardly be realised by internal employees or the company's own IT department, mainly for capacity reasons. If the aim is to provide comprehensive digital support for DM, in-house development should be avoided, and a commercially available system that can best map the company's own processes should be used.

Of these eight variants, three alternatives for the digitalisation of DM are often discussed in more detail: A solution from a combination of existing systems (depending on the focus of DM in the company, either on action management or key figures), an add-on that can be easily integrated into the IT landscape, and a dDM software (Figure 5.13).

These variants differ not only in terms of functionality but also in terms of cost. The task here is to select the right system – this requires a utility and profitability analysis. Customisable templates for both can be found with our templates (link at the start of the book).

The template for the requirements specification (see section 5.2) already contains the functionality to evaluate different variants in terms of their degree of fulfilment of the requirements and, thus, to determine their utility value. External dDM software generally achieves the highest utility value. However, it also generally has the highest (externally, quantifiable) costs.

The costs of implementing the software in the organisation and the savings made by digitising processes in DM can be estimated and used for a business decision, e.g., based on the amortisation period.

The template for this (link at the start of the book) is divided into six sections. Company data and other calculation values are entered in the first two sections. The other sections are calculated automatically (highlighted in grey). The annual financial benefit is shown first. This is compared with the initial investment and running costs. The amortisation period is determined from these values (Figure 5.14 and Figure 5.15).

Amortisation time estimate for dDM			
Company data for Areas with dDM		**Unit**	**Comment**
Number of teams	12		
Number of connected key figures	10		
Hourly rate engineer	90	€ / hour	
Hourly rate specialist	70	€ / hour	
Error costs	80.000	€ / year	
Personnel costs	12 million	€ / year	Alternative bases for estimating the savings - use only one
Profit		€ / year	
Calculation values		**Unit**	**Comment**
Duration of key figures preparation for meetings and Reports	0,5	Hours / Moderator	
Working days	220	Days / Year	
Printing costs, whiteboards and materials	100	€ / year and team	
Error cost reduction	10	%	
Increasing efficiency in the production	1	%	The lower the degree of maturity in the areas of knowledge management, data-based decisions, deviation management and problem solving, the higher the savings potential can be expected.
Software onboarding costs	15.000	€	The prices and calculation bases of different providers can differ a lot.
Software licence costs	150	€ / year and Team	
Procurement and installation of meeting equipment	2.000	€ / team	
Internal project expenditure PM	10	Days	May vary depending on company size and coordination effort.
Internal project expenditure IT	2	Days	May vary depending on operating mode and IT infrastructure.
Training costs	2	Hours / Person	
Operation and maintenance (Hardware)	100	€ / year and Team	
System maintenance (internal) PM	2	Days / year	
System maintenance (internal) IT	1	Days / year	

FIGURE 5.14
Template for calculating the amortisation period I – Imputed values. (Generated by authors.)

Benefit		**Unit**	**Comment**
Saving time in the preparation of meetings through automated key figures.	92.400	€ / year	
Elimination of printing costs, whiteboards and materials	1.200	€ / year	
Error cost reduction	8.000	€ / year	
Reduction in personnel costs	120.000	€ / year	
Profit increase	-	€ / year	
TOTAL	221.600	€ / year	
Costs - Invest		**Unit**	**Comment**
Software connection, customising	15.000	€	
Provide hardware	24.000	€	Number of teams can be reduced, if meeting corners are used several times or screens are already available.
Internal personnel costs	8.640	€	
Employee training in tool and possibly new processes	1.680	€	
TOTAL	49.320	€	
Costs - Ongoing		**Unit**	**Comment**
Licences	1.800	€ / year	
Operation and maintenance (Hardware)	1.200	€ / year	
Internal maintenance	2.160	€ / year	
TOTAL	5.160	€ / year	

Amortisation period		**Unit**	**Comment**
Amortisation period of the dDM	**0,228**	**years**	

FIGURE 5.15
Template for calculating the amortisation period II – Evaluation.

The savings from reducing working hours in implementing the DM can be estimated relatively easily based on effort and hourly rates. The elimination of printing costs and whiteboard material can be assumed as a lump sum. However, it is usually insignificant for the calculation. While these savings can be quantified easily, it is much more difficult to quantify the improvement of processes in DM itself and, above all, the benefits of increased DM effectiveness. It can be assumed that, for example, faster recognition of deviations, greater discipline in implementing the PDCA, and increased problem-solving and action quality have a lasting positive impact on production processes. However, it is almost impossible to find reliable figures for that. Nevertheless, arguments can be made about an increase in profits (labour or machine productivity) or a reduction in personnel costs of one to two per cent. A reduction in error costs can also be assumed.

The costs of system implementation and maintenance depend on the type of dDM system and the chosen provider. Hence, it must be enquired about and negotiated on a case-by-case basis. The prices and the calculation basis (users, teams, and plants) can vary greatly.

The amortisation period for introducing dDM is calculated depending on prices and assumed improvements. In general, it is well under one year.

5.3.3 Implementation of the System

The implementation of the dDM software follows the selection. This is divided into four steps, whereby steps 2 and 3 can run in parallel and vary in scope depending on the system selected:

1. Specification of organisational structure, users, key figures, and error categories
2. Setting up the system
3. Connection of data sources and neighbouring systems
4. Configuration of the dashboards

5.3.3.1 Specification of Organisational Structure, Users, Key Figures, and Error Categories

Firstly, the content of the dDM system is specified. The specification includes the organisational structure, users, key figures, and error categories.

Organisational Structure

Which part of the organisational structure should be mapped in dDM? Which regular communication and which shop floor meetings are to be digitalised? A precise specification of the teams to be digitalised is required – from the shop floor to management. The plant management, with its departmental or divisional management and staff units, is also a "team" of the organisational structure that must be considered. In addition, regular communication and shop floor meetings are specified, i.e., who meets with who and when, and what the agenda looks like. The result should include

- Teams (from the shop floor to management)
- Their hierarchical relationship to each other
- Regular communication and shop floor meetings incl., agenda

User

Which people will be involved in the dDM? What are their roles? In addition to the people in the previously specified teams, it is also important to specify other people to be involved. If, for example, the purchasing department itself will not be part of the dDM, it may still make sense to involve users from purchasing in the dDM in order to obtain information for the meetings and to be able to assign measures. If the dDM software supports roles, these must also be specified. A dDM team's role does not always correspond to the company's main role. The result of the user specification should include

- User (name, and e-mail address for notifications if applicable)
- Their team membership and, if applicable, roles in these teams.

Key figures

Which key figures are visualised on the dDM boards? How are these key figures calculated? What are their target and limit values? The key figures to be used are an important part of dDM and should be specified in detail. The most important properties of key figures are listed below, and an example is given.

Property	Example
Designation of the key figure	Quantity
Assignment to the organisational unit	Berlin – Production – Assembly – Line XGF
Target value	50
Lower limit value	45
Data source	Aggregation (sum)
Update interval	Shift

Depending on the scope of functions of the dDM, further advanced properties can be mapped, such as

- Linking the key figure with others for better context (machine availability and number of units)
- Time and user of last update for key figures

Error Categories

It is important to be able to analyse the deviations, measures, and problem solutions generated in the dDM. For this purpose, properties can be assigned to these elements, which are to be filled in by the users. These analysis categories can be used to create live analyses during operation with regard to reasons for errors, critical product groups, etc. Typical examples are

- Deviation category (safety, quality, costs, etc.)
- Fault categories (cleaning, setup, standstill, etc.)
- Product or product group
- Workstation or machine
- Customer or order number

5.3.3.2 Setting Up the System

After the specification, the dDM can be set up. This step is directly dependent on the chosen solution. Generally, supplied solutions are well-prepared for this phase and can be provided quickly (often automatically). If a dDM is created from scratch – an in-house development or a combination of existing software – setting up the software based on the specification can take considerably longer.

5.3.3.3 Connection of Data Sources and Neighbouring Systems

The data to be connected and their sources were specified in the specification. The fact that the dDM system can connect to the interfaces of the source system, in principle, has already been ensured *via* the specifications during system selection. Now, the connection must be established. Depending on the operating mode of the system and security guidelines (technical and organisational), different steps are necessary to varying degrees. Frequently required measures are as follows:

- Set up VPN connections between the servers of the source and dDM systems
- Make settings in the source system so that an export is permitted
- Make settings in the dDM system so that the data is loaded at the correct frequency and saved in the correct location
- Include business and parsing logic to aggregate useable KPIs from raw data

The exact steps are system-specific and should be agreed upon directly by the respective system managers.

5.3.3.4 Configuration of the Dashboards

Finally, the dashboards are configured. With the KPIs, suitable visualisations for conducting meetings can be implemented on dashboards.

5.3.4 M3: Operational System

After implementation, there is an operational system with linked key figures. The next task is to put it into use.

5.4 Ramp Up

For the system to be used successfully on a broad scale, it must be made available on-site, and users must be trained in using the software and, if necessary, in new processes.

5.4.1 Provision

The production areas need access to the dDM software. This requires screens with a PC and network access directly in the production environment or at the physical location of the DM meeting. Smartboards with touch functions are nice to have but not necessary; large screens with wireless keyboard and a mouse on a standing table have proven to be the most efficient means.

Comparably, a good network connection should be carefully ensured – long loading times understandably quickly lead to the system being rejected. This includes the end device's internet connection and the connections to neighbouring systems.

5.4.2 Training Courses

New elements in the DM, e.g., key figure definitions, problem-solving, or escalation processes, must be communicated to employees. For example, the software can provide support by displaying the exact definitions of key figures.

Employees must be trained in the functionalities so that they can use the software with confidence. Suppliers usually make training courses available for this purpose. Self-created solutions must be trained internally.

For larger companies, the amount of training required can be very high. For this reason, a dDM pilot should be used to train other experts to support the roll-out to other teams.

In order to organise the training courses as efficiently as possible, a role-specific concept has proven its worth. Typical roles in relation to software use are:

- The KeyUser configures and customises the dDM. They must, therefore, fully understand the system and be able to operate the system settings. Their training needs are greatest and can be extensive (approx. 1-3 days).

- Moderators need the skills to customise and conduct the shop floor meetings. They must be confident in the program's operation and know their scope for creativity (training time approx. 0.5 days). In addition, they must master and exemplify the leadership, moderation, and coaching routines described in this book to involve and motivate all participants appropriately and to anchor the digital DM process sustainably.
- The users must be enabled to record data, interpret key figures, recognise deviations, and record problem descriptions and solutions. The moderators usually do this based on specific cases.

The scope and time of the training also influence the acceptance of the system. If users are overwhelmed or left alone, they will be unable to use the system and will reject it. This includes the gradual introduction of new elements in the DM and the handling of the system itself. Technical affinity and expertise in dealing with digital systems positively affect the willingness to use dDM. Employees under 35 have grown up with digital systems (digital natives) and generally favour them. However, older employees are also familiar with digital systems, with the spread of smartphones since 2006. But there may still be team leaders or employees who are not confident using software systems. To ensure that they accept a dDM, they should be introduced to the system at an early stage [1]. Experience has shown that even these people appreciate the advantages of dDM after a longer familiarisation period. In some cases, they are even more enthusiastic than a digital native, as it represents a major innovation for them.

5.4.3 M4: Consistent Digital Daily Management

When the software goes live, the DM runs in a system that aggregates key figures, deviations and problems, as well as the associated corrective measures, consistently and logically along the meeting cascade and thus supports the DM control loop.

5.5 Roll-Out and Continuous Improvement

To achieve long-term success in improving deviation management through dDM, DM should be subject to continuous improvement. The software functionalities should be regularly reviewed and adapted.

5.5.1 Review of Target Achievement

Even after a complete roll-out, the dDM, like any other process, should be subject to a CIP. A standardised manoeuvre critique should already be part of every completed PDCA. The team should briefly ask itself how satisfied it is with the result of a PDCA (Has the problem been sustainably eliminated?) and how well the DM process and teamwork went (What went well? What can be improved?). If necessary, adjustments to the DM process and to the software can be derived from this.

Beyond this small improvement cycle, a regular (e.g., annual) review of the DM status should take place. To evaluate the processes in the dDM, the same questionnaire can be used that was also used to determine the status of the analogue DM (see Section 5.2). This makes it possible to determine the extent to which the DM has developed methodologically. However, the tool and its use should also be regularly reviewed.

The desired target status for the dDM is achieved when managers and employees routinely carry out their daily DM checks with the help of the software and use the available templates and documents. The PDCA is successfully run through regularly without further external impetus. In this state, managers and team leaders are freed from data procurement and preparation. They recognise deviations based on selected key figures that are automatically displayed and consistently linked across the different levels. All teams are aligned to the same strategic goals. In addition, employees and managers can use the time freed up for sustainable problem-solving and focus on implementing and following measures.

The status review serves to identify potentials in the individual areas, recognise good practices, and prioritise and initiate measures to improve dDM further. For example, it may be necessary to make methodological changes, redefine key figures, prioritise deviations differently, adapt problem-solving methods, or rethink the understanding of leadership. These changes must also be reflected in the software and may lead to adjustments by the supplier.

5.5.2 M5: Digital Daily Management as an Improvement Routine

"You have reached your goal!" – With dDM, the organisation can now make better decisions quicker and rectify problems more sustainably, thereby maintaining and expanding its market position even in times of change.

The established data basis also offers further potential through data analytics (see Chapter 7).

References

[1] Kandler, M.; Dierolf, L.; Bender, M. et al.: Daily management acceptance in global manufacturing. *Procedia CIRP* 115 (2022), pp. 190–195.
[2] Krips, D.: *Stakeholder management*. Berlin Heidelberg: Springer 2017.

6

Implementation of Digital Daily Management: Field Reports

ABSTRACT Numerous correlations and recommendations regarding dDM were presented. These are now illustrated using four practical examples with specific challenges, the selection of dDM systems, and implementation experiences. The companies did not have access to this book's findings, including the implementation method. The structure of the implementation method reflects their approach and experiences, which have been incorporated into the design of the method.

6.1 Production Data Acquisition and dDM in SMEs: Introduction at Filtration Group

This experience report was written by Volker Heuser, Head of Operations at the Öhringen site.

Filtration Group GmbH, headquartered in Öhringen, Baden Württemberg, is part of the Filtration Group Corporation, a US group Madison Industries subsidiary. Filtration Group GmbH has established itself in the global market with an enormous range of high-performance and innovative products for separating and filtering liquids, air pollution control, and process filtration. The plant management initiated the introduction of a dDM.

Annual turnover at the site: approx. € 100 million

Employees at the site: approx. 400

6.1.1 Preparation

In 2021, the Operations division set itself the goal of updating the analogue DM to improve deviation management and transparency. It was quickly understood that this goal would not be achieved to the expected extent with the analogue system. This led to the introduction of a digital version

DOI: 10.1201/9781003612681-6

to improve the company's key figures with better transparency, communication, and less effort. The Head of Operations took on the project management themselves.

A core team from Operations, Logistics, Quality, and IT was installed to ensure rapid implementation and an efficient start. This team completed the tasks quickly and consistently alongside all other normal tasks. This was the only way to keep to the schedule and cooperate with the service providers. Short control loops were helpful.

Stakeholders from other departments were not involved. Therefore, some interfaces cannot be operated, and other systems are still running in parallel. For example, measures in the quality area are still tracked in SAP so that quality can cover all areas of the company.

6.1.2 Analysis

The project's starting point was that key figures were not consistently defined, and there was a lot of work to archive them. Furthermore, there was a lack of discipline in conducting shop floor meetings in various areas.

The analysis of the IT landscape has shown that SAP and software for idea management, health/safety/environment, and training are the main neighbouring systems.

6.1.3 Design

The dDM was also intended to improve methodological expertise in analysing data. To this end, the meeting and KPI structure were completely reorganised, and new KPIs were assigned to each meeting. The biggest change was the addition of an OEE.

For this reason, the decision was made against the internal creation of the system from MS Teams.

With these requirements formulated in a specification sheet, various providers for dDM and MES were researched. Filtration Group decided to favour the Digital Teamboard from SFM Systems in combination with OEE. ai's production data acquisition. This made it possible to create the desired transparency of processes and improve them in guided meetings with binding deviation management.

Due to the complete reorganisation of the meetings and their content, the main challenge during implementation was to involve all departments. To achieve this, a tight schedule and budget were set and adhered to with discipline.

The approach of starting with a pilot helped overcome the initial hurdles easily and define the roll-out rules. Weak points were thus identified at an early stage and easily eliminated.

6.1.4 Ramp Up

The preparation time up to the go-live of the pilot was 6 weeks, and the roll-out was scheduled for approximately 3 months.

During this time, 10 hours of training was provided on how to use the tool (online and in person) and the necessary hardware was procured. It was possible to use a lot of the existing infrastructure and only add a few large screens.

After 3 months, a daily routine was switched to.
The installation went without any major difficulties, partly because both systems were hosted in the cloud. Only the connection of some SAP key figures became long-running, so the number of connected key figures was reduced. For example, some monthly key figures were not transferred as the effort was not worthwhile for a monthly view.

6.1.5 CIP

Further key figures were then implemented in a third phase, enabling the objectives of the DM reorganisation to be fully achieved: The transparency of the figures helped in day-to-day work, and the clear meeting structure has significantly improved communication.

6.1.6 Lessons Learnt from the Introduction at Filtration Group GmbH

In retrospect, the plant management at Filtration Group says that

- It is important to have a project manager who is focused on implementation and has a clear structure.
- It is important to define the key figures and their targets and limits clearly.
- It is important to build discipline across all areas and shifts.
- They are completely satisfied with the introduction.

This example confirms that methodically complete dDM software can be used to reorganise dDM to evaluate and sustainably improve processes with less effort. The introduction can also be implemented quickly with a unified staff between the Head and Operations and the project management.

6.2 dDM in a Large Corporation: Introduction at the Voith Group

This report was written by Julia Neubert, Project Leader of Operational Excellence.

The Voith Group is a global technology group. With its broad portfolio of systems, products, services, and digital applications, Voith sets standards in the energy, paper, raw materials, transport, and automotive markets. Founded in 1867, Voith is one of Europe's largest family-owned companies with around 21,000 employees in three divisions, sales of €4.8 billion, and locations in over 60 countries around the world.

Annual turnover: €4.8 billion

Number of employees: 21,000

6.2.1 Preparation

In 2019, the Voith Turbo Group Division decided to introduce a dDM and rolled it out at several locations in 2020–2022. The Division Paper started in 2022. The objective and background for the introduction of a digital DM tool is the implementation of a modern management system and the creation of transparency across hierarchical levels. As a standardised solution in all plants, the dDM intends to support managers with documentation (e.g., problem-solving), automated data preparation (no Excel lists and SAP exports), facilitate the tracking of long-term issues (problem-solving processes and success monitoring), and enable the creation of a knowledge base. In the long term, the knowledge database should enable users to learn from old mistakes and support AI applications in the sustainable solution of problems.

As part of the planning and preparation, various stakeholders had to be convinced and informed. This took place in various committees and regular meetings (e.g., operations committee, plant management group, IT project meeting, and works council committee) in the various specialist areas and hierarchical levels. In addition to involving the managers and employees of the plants, particular attention was paid to the stakeholders in the IT area (system connection and infrastructure) and the early involvement of the works council. The OPEX team was responsible for the overall project management and was supported by a dedicated IT project manager.

The project was received differently by different stakeholders, from great enthusiasm and anticipation of the opportunities of digitalisation to scepticism and fears. There were digitalisation fans who saw the advantages and opportunities of a digital solution, and there were the sceptics who questioned the added value, pointing out that "it already works well today". And finally, there were also the sceptics who saw too much effort for an unquantifiable benefit. In some cases, these could be reduced in advance through continuous communication, but in other cases, only in the course of implementation and after some time.

It has been shown that early involvement of IT is crucial, particularly in the case of a planned direct data connection.

6.2.2 Analysis

The existing DM processes were audited using a self-developed assessment sheet to prepare for the introduction. After approximately 10 years of analogue DM, an inhomogeneous picture emerged: many areas showed a good level of maturity, whereas individual rounds had become careless over time. For this reason, intensive on-site support at the DM boards and targeted method refreshers were planned for the roll-out, as an introduction without a broad basis in terms of methodological knowledge and the corresponding management culture did not seem promising. For Voith, a digital tool can only support the analogue process. Still, it can never compensate for a lack of culture and presence on the shop floor.

6.2.3 Design

As part of the preparations, the key figure cascade and the associated aggregation and escalation rules were discussed, clarified and, in some cases, redefined. This discussion helped to foster a common understanding of DM and to discover and resolve inconsistencies that had occurred over the years.

In addition to discussions on the standardised duration and time of shop floor meetings, processes and rules had to be tightened up (what happens when in the event of deviations). Technical conditions (automatic aggregation requires all data from the levels below) dictated and increased the regularity of the shop floor meetings.

For Voith, it was clear from the outset that they did not want to implement a self-programmed solution. Instead, Voith wanted to rely on a standard that was constantly being further developed and used by various companies. User-friendliness, simple system integration, and the continuous development of the tool were the key criteria for the decision. After testing several systems, the decision was made to favour the Digital Teamboard from SFM Systems.

The software installation was without difficulties. However, to make the changeover as smooth as possible, the DM corners were converted, that is, the hardware integration only took place after the software installation and configuration. The existing shop floor corners were predominantly equipped with 65-inch screens with a mouse and keyboard. A 65- or 75-inch touchscreen is used in a few shop floor corners.

The configuration also included automatically connecting key figures. It was important to carry out data validation, test calculations, and aggregation rules in advance, for example, when dealing with public holidays and spontaneous weekend shifts.

6.2.4 Ramp Up

During the Ramp up, a number of challenges had to be addressed. Convincing the "non-tech-savvy" required intensive support right from the start. Even though the system was easily customisable, the problem-solving templates looked different to what we were used to.

On the one hand, these topics were addressed through intensive on-site support, and the project management was on hand to provide users with advice and assistance. On the other hand, illustrated step-by-step instructions were provided for the changes; for example, in the problem-solving template, examples were used to train users in its use.

Voith used a two-stage training concept consisting of theory and practice. All key users and moderators were familiarised with the functions of the tool using the SFM Systems training platform (approximately 2–4 hours, depending on role). Secondly, the project team provided intensive on-site support for the DM meetings as part of the roll-out so that users could receive support at all times if they needed it. From the first test operation in pilot areas (approximately 4 weeks in 2 DM rounds) to the complete roll-out in all direct areas, the introduction took around 3 months per plant (excluding the preparation phase). Subsequently, minor adjustments were implemented, and individual DM rounds were connected in indirect areas. Due to the continuous involvement of users in the improvement and further development of the tool, the tool had a high level of acceptance and is firmly integrated into everyday life.

6.2.5 CIP

Some goals have not yet been fully achieved: In terms of transparency and user support, Voith is extremely satisfied and receives positive feedback from the organisation. However, it is difficult to quantify this, let alone the associated impact. In particular, the impact on problem-solving is difficult to quantify. In terms of knowledge management, Voith is satisfied with the current status and has thus laid the foundation for further applications. The "paper trail" and the associated risk of information loss are a thing of the past.

The organisation's feedback is used with SFM Systems to develop the software further.

6.2.6 Lessons Learnt from the Introduction at the Voith Group

Voith is very satisfied with the decision to support the DM with a digital tool. It has proven to be important for a successful implementation

- To provide intensive support for onboarding,
- So that any problems and ambiguities that arise are quickly resolved and
- Not lead to reservations or even a defensive attitude towards the dDM.

The smoother the system works from the start, the higher the level of acceptance.

Digital tools also force the organisation to clearly define rules, key figures, and processes. These necessary discussions are part of refining a common understanding. A well-understood DM methodology facilitates effective use.

Voith can recommend the implementation of a digital DM, but at the same time, must point out that it can only provide support and can in no way replace a targeted introduction of methods, a practised lean culture, and constant "sticking with it".

> In contrast to the much smaller introduction at the Filtration Group, the greater importance of stakeholder management is evident at Voith. This takes considerably more time in the Group structure during the preparation and design phase.

6.3 Digital Deviation Management in Large Companies: Introduction at Pilatus Aircraft Ltd.

This field report was written by Luca De Simoni, Product Owner.

Founded in 1939, the Swiss Pilatus Aircraft Ltd. develops and builds unique aircraft worldwide: from the legendary PC-12, the best-selling single-engine turboprop aircraft in its class, to the PC-7 MKX and PC-21, the associated simulators and the market-leading training systems for pilot training. The Pilatus team consists of over 2,500 dedicated employees, making the company headquartered in Stans one of the largest and most innovative employers in Central Switzerland. Independent subsidiaries in the USA and Australia are also part of the Pilatus Group. The company trains over 140 apprentices in various professions – promoting young professionals is a high priority. Pilatus has always consistently focused on Switzerland as a place to think and work and acts sustainably and environmentally consciously in all its activities. In particular, innovative approaches to increasing competitiveness are also promoted, and holistic solutions for production are developed.

Annual sales: CHF 1.347 billion (2022), EBIT 226 million (2022). The good market position allows for strong growth with the goal of expanding market leadership.

Number of employees: 2,599 or approximately 2300 FTE (2022), the majority of whom are based in Stans, Switzerland.

6.3.1 Preparation

The "Task Management System" project aimed to introduce a cross-departmental solution for creating, processing, and tracking tasks and measures in day-to-day business. Existing components, analogue methods, and new digital solutions were to be meaningfully combined.

The following stakeholders were involved in the project:

- IT: Definition of technical conditions and evaluation of possible solutions
- Operations leads (manufacturing, assembly, logistics, and supply chain): Project sponsorship and decision-making body
- Lean manager from the operations units: methodical input, as well as coordination with the general roll-out
- Team leaders and other managers: Selected team leaders to verify the implemented measures and to act as key users

To involve the stakeholders, various "mini-PoCs" were directly implemented to evaluate the user requirements, including Atlassian Trello, Atlassian Jira, hybrid shop floor, Microsoft Teams, and a dedicated digital shop floor solution. Subsequently, a concept for digitalising the shop floor was developed with the units (production, assembly, logistics, and supply chain) and verified in regular steering committees with the operations leads.

In some cases, very rapid adoption could be observed. In other areas, there was scepticism, but mainly about the methodology and much less about digitalisation. This can also be attributed to the fact that DM was still new at Pilatus. Both the methodology and the digital transformation were uncharted territory for many areas and stakeholders.

All key stakeholders were taken aboard. However, individual team leaders and their shop floor teams, which had been in place for some time, needed time to change and adapt. In some cases, the initiatives relating to DM had not been driven forward thoroughly.

On the one hand, a holistic understanding of DM was to be established at Pilatus, and on the other hand, digital tools were to be used to harmonise communication and task management. The following methodological and procedural objectives were defined:

- Networking has been established across the three business units of production, assembly, and supply chain ("communication at eye level").
- Shorter response times to problems.
- An established cross-divisional problem-solving process.
- Daily management standards have been defined and practised.
- Shop floor meetings have replaced production meetings.
- Cross-functional teams from different departments work on the shop floor (networked).
- Responsibilities are defined.

The following goals should be achieved, particularly with regard to digitalisation:

- Standardised hardware setup for shop floor teams.
- Information available on the shop floor in real time.
- Standardised KPIs are processed automatically and displayed digitally.
- Data-based decisions are made possible.

6.3.2 Analysis

The existing DM was analysed using a Pilatus tailored target/actual analysis focusing on "pain, gain, win" (problems, potential, already well implemented). In addition, root cause analyses were conducted on the main problems, and a detailed specification of the requirements was drawn up. In addition, an external partner was brought in to create a concept and define a roadmap.

Various core systems are already in use at Pilatus and require integration into the shop floor process. The following core systems were identified that are directly or indirectly part of DM at Pilatus:

- SAP
- Teamcenter
- Excel
- IoT machine data (Azure)
- Qlik Sense Dashboards
- Microsoft 365
- Other developed in-house applications

Not all systems were to be integrated directly into DM but rather incorporated into the shop floor methodology, depending on the situation. A major challenge was the very heterogeneous system landscape, with many systems disruptions in workflows, which was constantly being expanded due to Pilatus' long-standing production tradition and high level of specialisation. This resulted in many isolated solutions with sometimes conflicting principles. The lack of experience in DM made things even more difficult. However, Pilatus already added value in some areas with DM and thus anticipated an integrated, global system.

6.3.3 Design

For the first time, all DM principles had to be established and shaped by external references. The KPI structure was initially defined by each business unit's internal and external experts. The aim was to develop a standardised KPI framework with the five core target variables (safety, personnel, delivery, quality, and costs) to enable consistent KPIs across all three production units and the various shop floor levels.

With the help of the digital components, process flows and workflows within the DM methodology could be massively improved and applied on a scaling basis. The increased transparency through standardised metrics has clearly shown where there is a need for action.

Through a lively exchange with industry partners, many different systems were considered. These were compared with the Pilatus requirements and needs, and the selection was narrowed down. In the end, it turned out that the main decision had to be made first and foremost between make versus buy. As such, the shortlist consisted of

- Microsoft 365 (Teams, SharePoint, List, Planner)
- Atlassian products in interaction with existing components (e.g., JIRA)
- Dedicated daily management software

The following functionalities were ultimately decisive in the selection process:

- Existing platform solutions, some of which already offered shop floor functionalities
- Low-coding approach based on known technologies
- Reusability of existing methods and utilisation of existing resources

The decision was made in favour of a hybrid shop floor solution, using existing software (JIRA) and including local, in-house developments. This was brought to the shop floor *via* a standardised hardware framework. Due to the broad use of JIRA in engineering, customer interaction, tracking of logistics information, and project management, the shop floor problems and measures could be ideally integrated.

One of the challenges during implementation was finding suitable hardware for user interaction with the software. This had to conform to the Pilatus requirements and achieve the organisation's best possible penetration. In the business units, initial areas were supported by those responsible for DM and then continuously expanded as a blueprint.

One positive aspect to emphasise during implementation was the cooperation within the team. The team included various business units, so the project benefited from the expertise of the different departments. Together, they successfully defined and realised the implementation roadmap. The progress and implementation risks were discussed in regular consultations between IT, business, and external partners. Some points on the roadmap could be used for subsequent initiatives to lay the foundations for core technologies such as a data warehouse or an IoT platform.

Finally, the existing workflows were modified to facilitate successful digital task management, which business units also accepted.

Full high-definition touchscreens of 55″ were selected as the hardware and combined with Intel Mini PCs. Digital signage software was used on these modified Windows 10 machines to integrate web applications. This meant that the digital task management system could be easily integrated into the shop floor meetings and, together with the digital signage solution, further information, such as company news, could also be communicated.

6.3.4 Ramp Up

After a few weeks, the pilot area was set up. Subsequent training for a division on digital DM took around 8 months, and the broad roll-out within the organisation took several years. In particular, it took time to develop an understanding of the shop floor methodology and anchor it in the organisation.

One important requirement was to make user authentication on the screens as simple as possible. This was difficult at the beginning due to internal policies and system restrictions. It was resolved through the consistent implementation of single sign-on and the assignment of users to the shop floor. Due to the ongoing roll-out, licensing new employees was also a challenge. In addition, the various other systems meant that different authorisations were required, each of which had to be triggered in a coordinated manner to obtain full access to all information.

The project encountered a very heterogeneous system landscape when connecting data sources and key figures. This resulted in more work than originally expected, especially at the beginning.

Inconsistent process definitions made it difficult to implement digital task management, especially the harmonisation of workflows. Even existing dashboards could not simply be used for the newly created shop floor levels – in the process before digitalisation, company-wide basic requirements should generally be created to shape the transformation step-by-step and not as a big jolt.

6.3.5 CIP

Ultimately, many of the goals were achieved by the end of the project. For example, a cross-departmental task management system was established to manage tasks, measures, and suggestions for improvement digitally. This system ensures a continuous flow of information from the line all the way up to management and units outside of Operations.

The project succeeded in harmonising and standardising the workflow structure. Tasks are recorded and processed in the same way in all implemented areas. The digital task management system is used in all areas *via* standardised hardware. At the same time, a flexible visualisation solution was created for all web applications. Due to individual maturity levels and specific requirements, the hybrid shop floor allows flexible levels of digitalisation and rapid onboarding through interaction with digital components.

A fully digital DM solution across the entire Pilatus organisation has not yet been achieved in the project. A final decision in favour of such a solution is still pending. There is also still a lack of a standardised, digital key performance indicator framework used consistently from the base to top management. To achieve this, all master data in the core systems would first have to be cleansed.

6.3.6 Lessons Learnt from the Introduction at Pilatus

Successful implementation of a digital task management system requires central project coordination. This is especially true when the organisation is divided into several business units for production, assembly, and logistics, as is the case at Pilatus. Because the foundations were only laid with digitalisation, the roadmap and the speed of implementation also had to be regularly reassessed. In addition, the north star in the form of a central KPI framework should be known at the beginning because the objectives depend on it to a large extent.

Introducing dDM is a change process that can take a long time, especially if there is little basis for DM. It requires a high level of consistency from top management to employees. Existing IT system landscapes and customer benefits should be sensibly integrated and only gradually replaced.

In addition, management attention must be emphasised from the outset. The importance of the project must be recognised and accepted by the management. The strategic cornerstones and direction of the project must be defined in advance and reviewed regularly. For a complete digital transformation, methodological expertise should already be established, and data sources and neighbouring systems should be harmonised. Only then can the hybrid shop floor be sensibly replaced digitally.

In contrast to the first two examples, Pilatus did not introduce dDM software in this area but initially introduced digital task management. This enabled some goals to be achieved, but licence and integration costs were also incurred. However, without the KPI framework and its integration into DM, the important data basis for fact-based prioritisation is missing. Furthermore, without the control of top management, it has not yet been possible to harmonise the units with each other.

6.4 Digital Deviation Management in SMEs: Introduction at Munsch Chemie-Pumpen GmbH

This field report was written by Stefan Munsch, Managing Director, and Oliver Herz, Head of Application Development.

The family-run company Munsch Chemie-Pumpen GmbH specialises in pumping highly corrosive and aggressive liquids. Safe pumping for people and the environment while being energy efficient are the company's core competencies. The special pumps are customised and come in a wide range of variants. With an export share of 80%, specific customer requirements and different regional standards must be met. In order to improve the current process of deviation management and to achieve sustainable problem-solving in the future, the company decided to use a digital deviation reporting system instead of a paper-based reporting card.

Annual turnover: €25 million

Number of employees: 125

6.4.1 Preparation

Innovation has always been a driver for the further development of Munsch, and digitalisation is a key enabler in strengthening the company's ability to innovate. As part of the company's digital transformation, replacing the traditional paper-based deviation reporting process with a digital system was essential. The aim was to improve the existing deviation management and establish a sustainable problem-solving process. Firstly, Munsch wanted to gain a more comprehensive overview of the deviations so that it could focus on the most important problems and invest time in analysing the causes. In this way, long-term measures could be defined to prevent the recurrence of similar problems. Historical measures were documented, and a knowledge database was gradually built up, which is useful for eliminating future deviations.

Almost all relevant stakeholders were involved during the preparation process. In the beginning, it was necessary to talk to all employees and managers from production throughout the entire production department who were responsible for processing and creating the report cards. This made it possible to discover what was happening in the reporting process and what challenges existed. The IT department also played an important role in this project, as it had a good overview of the relevant systems and information flows. In addition, the company's management also played a major role in motivating employees to drive the digitalisation transformation forward. The involvement of stakeholders from different areas varied depending on the area. The decisive factor was the extent to which employees were familiar with the importance of good deviation management and problem-solving processes.

6.4.2 Analysis

The company's original deviation reporting system was based on paper. There were several problems with the analogue process:

- The creators of the cards did not receive any appropriate feedback after the deviations had been rectified. This impaired their motivation to create further report cards.
- The paper-based report cards were easily lost during transport. The processing authorisations for the cards were difficult to manage.
- The management level could barely gain an overview of the deviations that had recently occurred. This made it difficult to plan resources for a systematic problem-solving process.
- There was no standard documentation for the report cards. Therefore, resolving the new deviations using the measures already contained in the report cards was difficult.

After the analysis, the conviction grew that digitalisation is a suitable way to achieve the set goals because it can eliminate the problems mentioned above. The digital deviation report card should be documented in the system, assigned to those responsible or passed on to various departments. The progress of the card, such as new adjustments and feedback, could be checked at any time. The data could be used to develop a dashboard for the management level, making it possible to track the deviations and even cluster the deviations from the past.

6.4.3 Design

Munsch already had standardised deviation management processes and PSP processes that had been optimised for many years. Employees at all levels are also very familiar with these processes. Therefore, after an analysis with the various stakeholders, the existing processes were not changed, but the previous analogue processes were digitally replicated to increase the transparency of deviation management. The most important requirements for the digital system were defined as follows:

- Employees should be able to record deviations directly during the production process. The information stored on the deviation report card should contain the master data from the ERP system, such as order and operation, so that more time can be gained for implementing the measure.
- Ease of use, interoperability, and no system discontinuity.
- Costs: Regardless of whether it was "make or buy", the effort had to be made to analyse the entire process, involve the stakeholders, define the key figures, and program the interfaces. So, the only question was whether developing the software ourselves was cheaper than buying ready-made software.

Ultimately, there were two implementation options: A commercial product from a third-party provider or developing the system in-house. After weighing up the options, Munsch concluded that developing its own system would better fulfil the three requirements mentioned above. Therefore, it was decided to integrate a specially developed digital deviation reporting system into our existing ERP system.

The entire implementation process went very smoothly. It took two weeks to integrate the designed digital deviation reporting system into the existing ERP system.

6.4.4 Ramp Up

The IT department provided training for the system, and the development staff were responsible for the training. It took 8–12 weeks for the system to be used in daily production without errors. During this time, system errors were always rectified at short notice (within a week), and feedback from employees was collected. Based on the feedback, the system was gradually adapted to achieve more efficient utilisation. The only difficulty was that the employees had to get used to the new digital workflow. After switching from analogue to digital deviation report cards, the entire process from recording to processing was carried out in the system, which was very different from before. The good thing was that employee acceptance was very high, as the employees were already familiar with the front end of the ERP system. This is why they were well-integrated into the project from the outset.

A powerful BI dashboard was redesigned for the management level to track the status of the digital deviation report cards. The dashboard displays the status of all report cards in visual diagrams and allows them to be filtered by categories such as area or date. In addition, some key figures have also been defined, such as open cards in an area. If the open cards exceed the defined limit, the person responsible for the area must take prompt action.

No additional hardware was purchased during the introduction. The digital deviation reporting process requires terminals at each production process, which employees can use to fill out the digital deviation reporting card directly during the production process. As the terminals are already in place, no new hardware was purchased for this project.

6.4.5 CIP

Munsch is extremely satisfied with the digital deviation reporting system, as its use has resulted in better transparency regarding deviations from the shop floor to management. From now on, deviations on the shop floor are documented directly by an employee in the reporting system. The digital deviation report cards are transmitted to the management level, where responsibilities can be assigned. The BI dashboards also enable deviations that have occurred to be analysed. In the next step, this system will also be rolled out in other interface departments to integrate further deviations. With a better database, analysis functions can be developed in the medium term to support employees and managers in their decision-making.

6.4.6 Lessons Learnt from the Introduction at Munsch

To summarise, the following points are important when implementing the digital deviation reporting system:

- Gemba: The digital transformation of the analogue process cannot be decoupled from the "location of the event". This is why deviations should be recorded directly during the production process. It is also important that the card creators receive feedback on their report cards to increase their motivation.
- Communication: Good communication can accelerate the digital transformation. For example, exchanging good practices between departments can help ensure that people are better supported during the introduction.

- Leadership: The digital transformation must be driven by managers from the top down. Team leaders and top management should act as role models to drive the use of the system and gain employee acceptance.
- Mindset: Digitalisation is only one tool for process optimisation; the core must always be the mindset. Therefore, increasing employees' understanding of deviation management and building a culture of error in the team is always the most important factor in achieving a sustainable problem-solving process.

The example shows that even small companies can benefit from digitally supported improvements in internal communication.

7

Assistance Systems in Digital Daily Management

ABSTRACT Once dDM has been introduced, new potentials emerge. In particular, insights and recommendations for action can be generated for users by analysing available data. This chapter presents tried-and-tested data mining use cases to support dDM. The focus is on how information for decision-making can be extracted from performance and text data using data analysis methods [1].

7.1 Data Analytics to Identify Correlations in KPI Networks

The relationships between the key figures of different areas and levels are complex and often difficult for the observer to understand. The easier it is to estimate the effects of a change in one key figure on other key figures, the earlier and better countermeasures can be taken.

7.1.1 Initial Situation

Many KPIs often lead to information overload for employees and managers, as the relationships and dependencies between the various KPIs are difficult to understand [2]. Although the KPI system should be kept simple (see Section 2.2), aggregating different KPIs into one overarching KPI is unavoidable in large organisations. Due to this complexity, changes in KPI trends and correlations between different KPIs are no longer intuitively comprehensible or even predictable. This results in a great demand for system-based decision support [3], simplifying the evaluation of a KPI system.

7.1.2 Solution Approach

In a digital DM, there is a comprehensive database of key figures and their trends. Until now, only deviations of these key figures from targets have been used for reactive improvement. However, time often plays a decisive role. The sooner managers know what challenges they can expect, the sooner they

 DOI: 10.1201/9781003612681-7

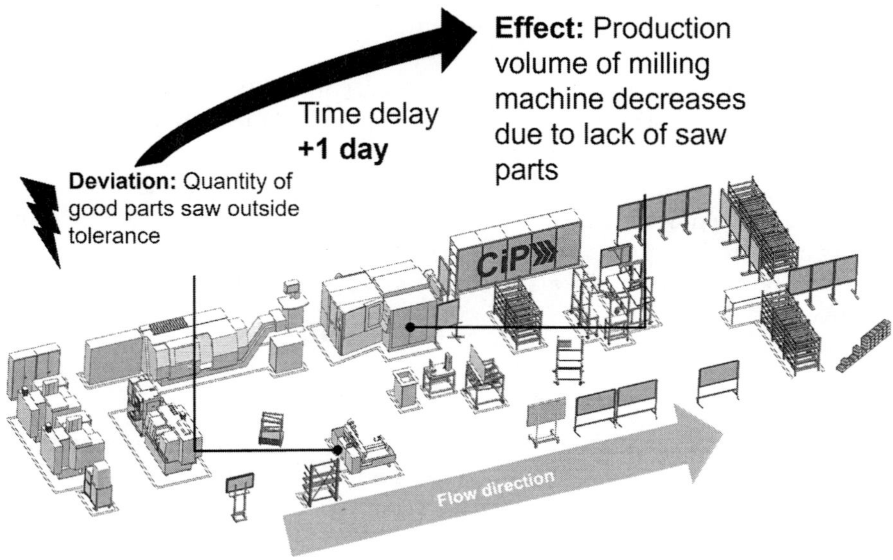

FIGURE 7.1
Time delay between two processes [Example from CIP process learning factory].

can take appropriate measures. For example, if quality problems have arisen in upstream processes, it is highly likely that these will also become noticeable in their own area in the next few hours or days. Data analytics can be used to quantify these correlations based on historical data and thus make them usable for future situations. For example, it is possible to find out how long the time lag is between problems on a system and the impact on subsequent processes (Figure 7.1).

The aim is, therefore, to find correlations and dependencies in the key figure network. Statistical measures of association are suitable for this purpose. They can indicate the degree of dependency or independence between two KPIs. If associations between KPIs are recognised, they provide information about relationships in the value stream. The correlation is a widespread representative of these association measures, which offers certain advantages due to its good interpretability. Correlation analyses can be used to recognise cross-process and cross-hierarchy relationships and thus prevent silo thinking: for example, a problem on a production line will have an impact on meeting deadlines in final assembly and, thus, ultimately, on the plant's delivery reliability.

The recognised correlations can then be used to manage production processes or the entire company proactively. In addition, KPI systems can be simplified by dispensing with KPIs that correlate strongly with others (i.e., they have the same or a very similar trend).

There are two particular difficulties in the technical realisation of such decision support:

Firstly, the type of correlation (e.g., linear and quadratic) between different key figures is not known in advance, making it difficult to select the parameters for the analysis.

Secondly, there is often a time lag between the causative and the dependent key figure in the production process. As described above, quality problems in upstream process steps often only become noticeable in their own area after hours or days. Accordingly, a system must be able to determine this offset iteratively.

7.1.3 Experience

The extent of the benefit of analysing correlations in KPIs essentially depends on which correlations can be identified and how strong they are. Extensive tests on real practical data sets provide two interesting results: Very strong correlations were often found between key figures. If key figures measure the same thing, the key figure system can be simplified. Secondly, interesting time-delayed cause-and-effect relationships between key figures were identified. This made it possible to determine how long problems in one area take to appear in subsequent areas and which performance parameters they affect. This makes it possible to implement early warning systems.

To visualise the results, it is appropriate to display the identified relationships in a knowledge graph (Figure 7.2). This graph shows the results of time series analyses of different key figures. The correlation coefficient shows how far apart the values of two key figures are from their own mean value at a certain

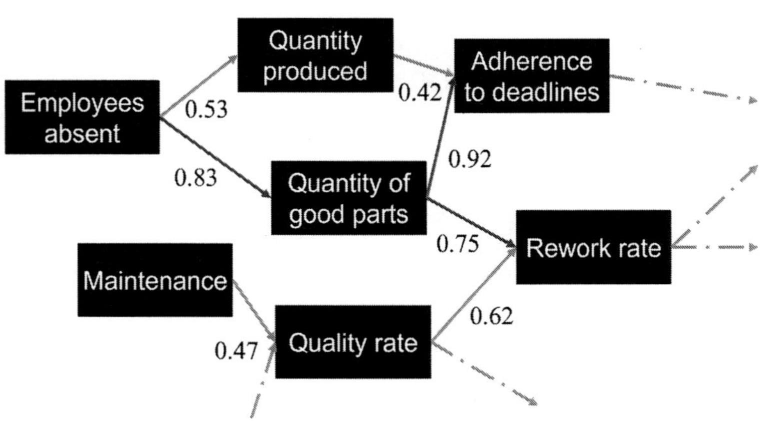

FIGURE 7.2
Recognised correlations between key figures.

point in time *t*. If they move away from their mean value at the same time and in the same direction, this is referred to as a positive correlation. If the deviations from the respective mean value move in opposite directions, they are negatively correlated. The strength of the correlation fluctuates between –1 and 1 as the deviations are normalised.

Although there is no generally valid statement as to when correlations are considered strong, medium, or weak, the following guidelines are often used [4]:

- Correlation coefficients between –1 and –0.7 (or 1 and 0.7) are considered strong correlations.
- Correlation coefficients between –0.7 and –0.3 (or 0.3 and 0.7) are considered medium correlations.
- Correlation coefficients between –0.3 and 0.3 are considered weak correlations.

The direction of the relationship, that is, the connection between cause and effect, is indicated by the respective arrow. This provides a clear overview of existing relationships in the key figure network.

The practical tests also show the importance of data quality for successfully using such a system. Almost all the data analysed showed a large number of missing and/or incorrect KPI values. This makes it difficult or completely impossible to calculate the correlations. In addition, it is clear that the existing KPI network also plays a major role in the final results. In order to realise the full potential of such a system, the selection and complete and consistent recording of the KPIs must be well-prepared, and good data quality must be ensured.

The correlation analysis makes it possible to improve the dDM in three steps:

- Creation of a sufficiently large and well-maintained network of key figures: The software used should collect data as automatically as possible or request it on a mandatory basis and store it over time.
- Generation and interpretation of a knowledge graph: Answering the questions of which key figures are redundant and which have a relevant time-delayed relationship.
- Implementing the findings in everyday life: using the recognised correlations to simplify the KPI systems and install early warning systems

7.2 Knowledge Management through Recommendation Systems

The other data analytics use cases deal with text data from deviation management, in which knowledge is documented throughout the process. The first approach shows how recommended systems can use this text data for low-effort knowledge management.

7.2.1 Initial Situation

When a deviation occurs, workers or managers need to analyse the root causes on time and take appropriate measures to resolve the deviation and eliminate its adverse effects. The longer the response time after a deviation, the more value is lost to the manufacturers. In this case, workers or managers need a lot of relevant information to support their decision-making. As such, this information gap can be eliminated through using recommendation systems. Recent advances in large language models (LLMs), such as ChatGPT, have demonstrated superior learning and reasoning capabilities, making it ideally suited for enhancing recommendation systems for solving complex tasks [5]. However, implementing this technology in domain-specific areas, such as production-specific scenarios, still needs to be investigated and defined.

7.2.2 Solution Approach

The potential applications of LLMs in deviation management are illustrated in Figure 7.3. Deviations on the shop floor are primarily identified through KPIs, audits, and Andon systems. From a system perspective, the outputs of Andon and KPI systems are typically Boolean or numerical data, which enable the identification of deviations using rule-based algorithms. LLMs can assist users in analysing data from these systems based on predefined rules, enabling more precise anomaly detection. In deviation analysis, employees and managers rely on their expertise to diagnose root causes. LLMs, enhanced with cue engineering, can replicate human reasoning patterns by breaking down complex problems into sequential logical steps. By integrating domain-specific knowledge into LLMs, these systems can be tailored to comprehend the unique dynamics of the shop floor. When deviations occur, LLMs can aid in structuring deviation analysis and suggesting possible causes, thereby accelerating the process and mitigating information bias. Once the root cause is identified, the plan-do-check-act (PDCA) cycle can be executed, with LLMs assisting in drafting actionable plans. During the Act phase, LLMs can function as an operational guidance system, extracting relevant procedural knowledge and providing

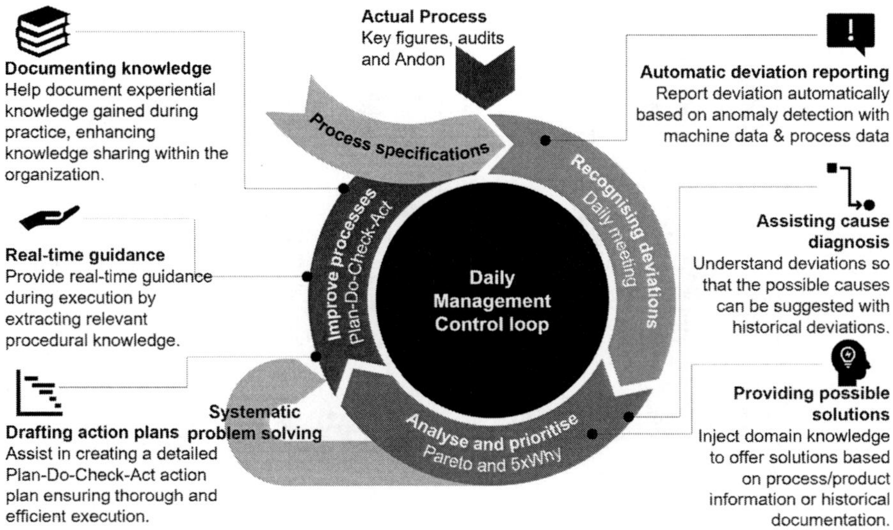

FIGURE 7.3
Applications of LLMs in DM [6].

real-time support. Additionally, LLMs can facilitate the documentation of practical insights and enhance knowledge sharing across the organisation.

7.2.3 Experience

In the shop floor application scenario, given the involvement of a specific internal domain knowledge base, retrieval-augmented generation (RAG) is suitable for generating relevant answers accurately and efficiently in the recommender system. RAG combines pre-trained parametric memory with non-parametric memory for language generation [7]. The system design is shown in Figure 7.4.

The deviation analysis process relies on three key knowledge areas: production process information, standard operation procedures (SOPs), and historical deviation records. Production information covers value streams, while SOPs detail station and machine operations. Historical records document past deviations and their resolutions. These data, often in varied formats like text, tables, and images, require efficient processing for accurate interpretation. To address this, RAG-based methods extract information using LLMs or rule-based approaches, converting heterogeneous data into standardised text descriptions. These descriptions are chunked, embedded, and stored in a vector database with indexing for traceability. Users can interact *via* text queries or predefined prompts. The retriever processes inputs, searches the database, and then LLMs generate responses.

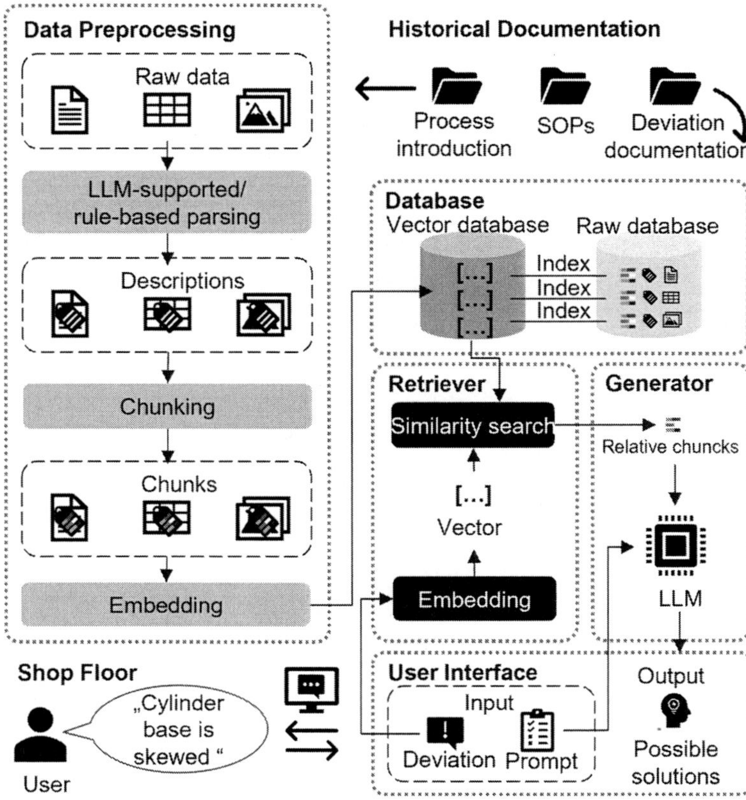

FIGURE 7.4
RAG pipeline on the shop floor [6].

As shown in Figure 7.5, to use the recommender system, the workers or managers on the shop floor should define the deviation statements in the system. After submitting the input, the system generates the possible root cause and solution and provides sources.

Testing showed the system handles heterogeneous data well and provides useful information to workers for decision-making. The implementation of such an LLM-based recommender system should also consider the following:

- The decision to use open-source or closed-source LLMs is based on data security, application scenarios, and input costs.
- The construction of the knowledge base has a very strong impact on the system's performance, and this process should be completed before applying the LLM.
- In order to lower the barrier for workers to use the system, it is better to combine the system with existing ones, such as the dDM.

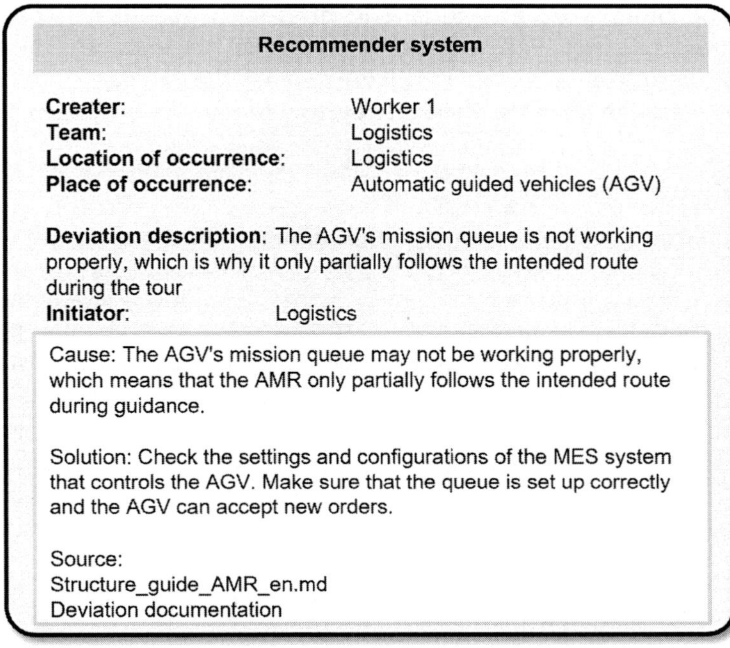

FIGURE 7.5
RAG in practice.

7.3 Knowledge Graph for Structuring the Deviation-Related Documentation

The deviation-related documentation always includes descriptions of possible causes and countermeasures that are valuable for deviation management and problem-solving. This use case shows how to extract the knowledge contained in the deviation documentation and visualise the relationships among the heterogeneous data in a knowledge graph.

7.3.1 Initial Situation

The documentation of deviations connects numerous entities involved in the production process, including people, processes, organisations, machines, and products. However, the complexity of the documentation also causes challenges for its practical reuse. Descriptions of deviations are difficult to standardise, so there is a high probability of ambiguity in understanding the text. The textual data can also contain errors, making it challenging to accurately process and extract valuable insights (see Section 7.6). Therefore, extracting representations of the relationships between deviations, measures, causes, and other relevant information from such documents containing unstructured data is challenging. Without this relational structure, the person accessing the information would have to search the records item by item using keywords, making it difficult to reuse the data.

7.3.2 Solution Approach

As a tool for organising and integrating different knowledge and information, the knowledge graph (KG) has been a focus of research in the computer field since it was introduced by Google in 2012. It can trace knowledge and reasoning, significantly assisting decision support across various domains [8]. A KG offers a structured and interconnected representation of information, enabling better decision-making, knowledge reuse, and the development of advanced recommender systems to drive continuous improvement and efficiency in the production process. It can support DM by providing possible measures, clustering the recurring deviations, visualising the problem–cause–solution loop, and more application scenarios. A top-down KG construction process contains five main tasks. *Data acquisition* describes the data collection required for KG. *Ontology construction* is used to build a semantic explanation of the domain knowledge [9]. *Knowledge extraction* mainly targets semi-structured and unstructured data. Its purpose is to discover valuable information within unstructured data and extract it. *Knowledge fusion* mainly focuses on the integration of different data. *KG construction* involves integrating all the data based on a predefined ontology structure into a graph-based database, which can be visualised in a graphical format.

7.3.3 Experience

An ontology structure is constructed based on the data acquisition and understanding of the example production, as shown in Figure 7.6. This ontology structure is centred on the occurrence of deviations. It contains four types of core content: information related to the classification of deviations, information on the location where the deviation occurred, information on the product where the deviation was found, and information related to the deviation event. The employees record deviation event data in the form of unstructured text in the deviation documents, which need to be extracted into the KG through knowledge extraction.

Knowledge extraction focuses on the unstructured data in the deviation documentation, that is, the descriptions of symptoms, causes, and measures. Based on the analysis of the writing characteristics, three types of node entities are set up: Object, Status, and Action, where Status and Object can describe symptoms and causes. The measure needs to be described through Action and Object. Four relationship entities are set as is_treated_with, appears_

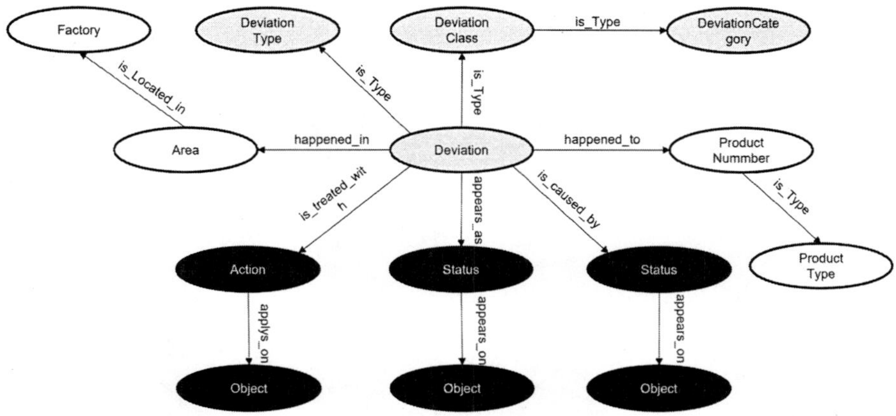

FIGURE 7.6
Ontology structure for knowledge graphs in dDM [10].

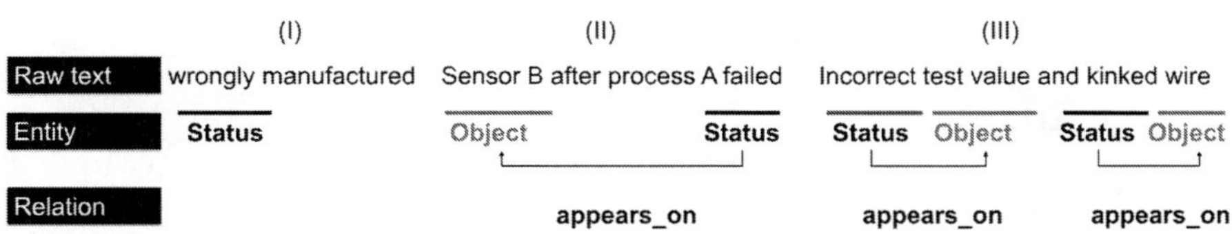

FIGURE 7.7
Example of relations between entities in shop floor data.

as, is_caused_by, and appears_on. For entity extraction, a special Natural Language Processing (NLP) called named entity recognition is used to automatically recognise the object, status, and action as well as their locations in the documentation, which looks like as follows:

[„Plug glued\r",{"entities":[[0,7,"OBJECT"],[8,16,"STATUS"]]}]

where "Plug" is recognised as an object with a position from 0 to 7 and "glued" is recognised as a status with a position from 8 to 16.

Rule-based relation extraction methods are suitable for text-poor linguistic environments [11]. By analysing the linguistic patterns in the data, three typical relationships are summarised, as shown in Figure 7.7.

The main task in the knowledge fusion process is to connect structured and unstructured data. Finally, using the py2neo toolkit, data is stored in the Neo4j graph database, and the visualisation results in the following graph, as shown in Figure 7.8.

With the KG, the deviation-related information can be stored in a graph-based database. In addition to storing the information in KG in a structured way, the KG from deviation documentation can achieve more convenient

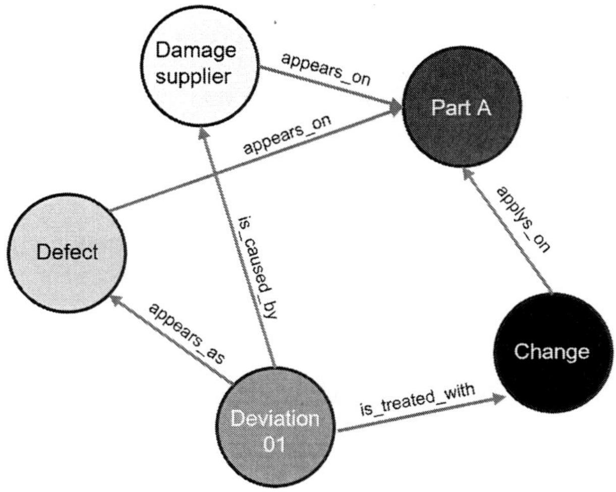

FIGURE 7.8
Knowledge graph based on shop floor data.

query and statistic functions. For example, if workers on the shop floor have a deviation relevant to "Part A" and "defect", they can use the KG to find possible root causes and measures from the past. Based on the KG, more use cases can be achieved, such as:

- Find the deviation clusters for certain objects.
- Summarising the countermeasure for solving root causes.
- Forecasting the potential impact of new deviations.

The KG can also be combined with the RAG system described in Section 7.2 to improve the knowledge retrieval performance of LLM-assisted recommendation systems.

7.4 In-line Prioritisation of Deviations through Data Analysis

A graph-based analysis is well suited to carrying our retrospective analysis and initiating problem-solving processes. By incorporating additional data, prioritisation can also provide direct benefits in everyday life.

7.4.1 Initial Situation

In the daily DM process, a problem-solving process is not started for every deviation in order to determine its root cause, as problem-solving processes can be costly and time-consuming. They should, therefore, be started with caution, depending on the priority of a particular deviation [12]. When prioritising and deciding on a problem-solving process, it can be helpful to answer the question: "Does eliminating this particular deviation contribute to our long-term goals, or is another problem more important?" When prioritising according to the Pareto principle, the question is asked: "Which problems are responsible for most of the deviations from the objectives?" (see Section 3.4). However, existing approaches to prioritisation are usually subjective or only consider one aspect for prioritisation, for example, the frequency of occurrence or the severity of a deviation. In many cases, prioritisation is decided based on personal judgement and experience [13].

7.4.2 Solution Approach

On the one hand, production generates data that describes a deviation. In many companies, this data is free text entries. The challenge lies in recognising the same error patterns and assigning them to each other so that, for example, clusters can be identified. Using conventional methods, it is very time-consuming to analyse free text entries to find similar deviations from the past and categorise them. The NLP offers help here, as it can examine different text data for similarities and thus offer suitable suggestions for a new entry.

In addition, a lot of data can contribute to prioritising a deviation: the affected product variants or machines, the duration of the deviation, or the costs incurred for rework. This process and problem-related data can be collected and documented by the dDM system from various sources (such as production data acquisition or the Manufacturing Execution System) and then made available for automatic data analysis [1].

Only relevant data should be selected for the prioritisation of deviations. As the prioritisation of deviations should contribute to achieving the overall objectives, the relevant data should include influences on the organisation's overall objectives, for example, optimising delivery reliability requires time totals for each process, and reducing rework requires historical rework records. This data and its weighting should be determined with the help of experienced employees and managers. By summarising and standardising the current data values, individual factors (such as cost or time factors) can now be determined. This way, factor values can be calculated for the deviation categories and combined into an overall priority ("error score", Figure 7.9) [14].

The error score for each deviation is determined by a combination of the trend of the increase in this deviation over the last period and the cost of rectifying it. The trend and cost can be calculated from the data stored in the dDM system. The factors are multiplied by specific weights and added together to obtain the defect score. The higher the error score, the higher the prioritisation. The weights should be constantly updated, as the importance of the various factors can change over time [15].

Prio	Prioritisation list	Tenence evaluation	Cost evaluation	Error evaluation
1	Damaged screws	0,669	0,856	**0,800**
2	Missing part A	0,722	0,782	**0,764**
3	Slot torque not OK	0,584	0,474	**0,507**
4	Missing part B	0,942	0,152	**0,389**
5	Tool damaged	0	0,455	**0,319**
	...	...		

FIGURE 7.9
Prioritisation list based on error score.

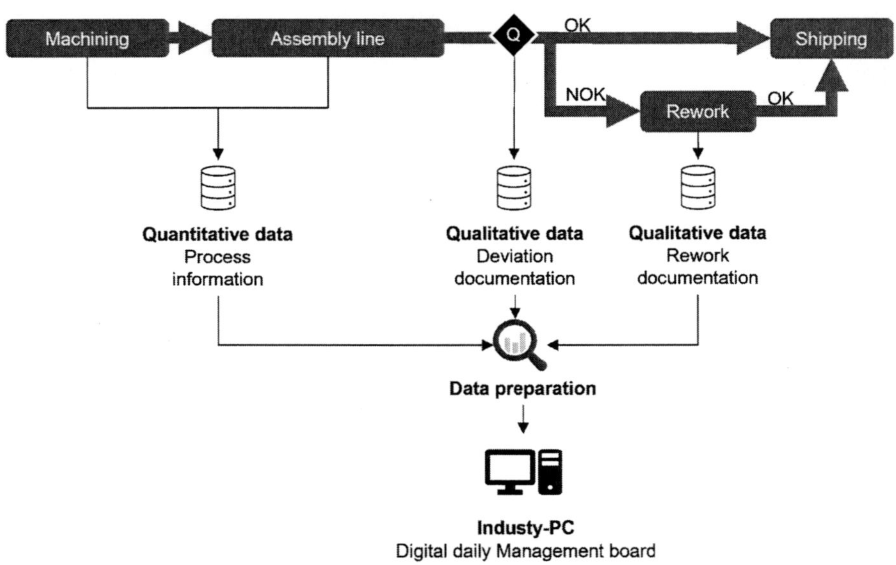

FIGURE 7.10
Implementation of error scores in the rework process.

7.4.3 Experience

The procedure described above was applied to data from an automotive company's production line. The aim was to provide the DM team with a list of prioritised deviation based on the fault scores as a decision-making aid.

As shown in Figure 7.10, a visual inspection and quality control is carried out at the end of the production line using testing devices. The deviations are recorded *via* terminals at the quality gate and are available in the dDM system. Products with deviations are reworked. Once the rework has been completed, the employees record their rework activities and the time required in the dDM system. Before the error scores were introduced, the DM team reviewed the rework logs daily and identified the most serious deviations to take countermeasures and adjust the personnel planning for rework. With the defect scores now available, the prioritisation of quality deviations was greatly improved. Applying the prioritisation method to historical data showed a different prioritisation than the one carried out by the DM team in 35% of cases. More targeted action and personnel planning have led to a significant reduction (over 20%) in rework time since the go-live in September 2021.

This example shows that data-based decision support can be used very effectively and practically to prioritise deviations.

It will be introduced in a similar way to the text-based assistance systems already described:

- Pre-processing of data relevant to the description and categorisation of deviations (see Section 7.6)
- Definition of influencing factors for prioritising deviations with management experience employees
- Weighting of the factors and calculation of the factor values
- Calculation of the error scores and creation of the prioritisation list

7.5 Chat Mining for Structuring Informal Problem-Solving

The exchange of information on operational topics and, therefore, deviations often takes place on chat platforms. This use case shows how structured data for prioritising deviations or solving problems can be generated through this informal exchange of information.

7.5.1 Initial Situation

Even with a very well-functioning dDM, there is still a latency period from when a deviation is reported until it is dealt with within the DM control loop or until it is escalated to the next management level. Due to the high pressure to act, the structured process is often skipped, and employees and managers are approached directly. Although a quick solution is often found and operations can proceed, this approach has two disadvantages: Success depends on individuals who need to know the right people to contact and the knowledge about a problem and its solution remains only with the people involved. No analysable data is generated.

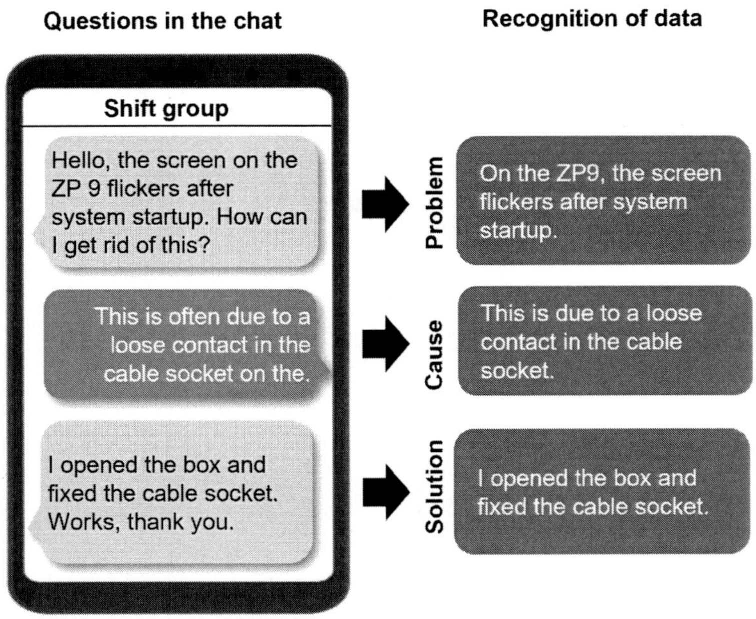

FIGURE 7.11
Recognising relevant content in chats [16].

7.5.2 Solution Approach

In many production areas, informal chat groups exist. These chats are used, for example, to swap shifts, exchange private information, and ask for help in the event of errors. The information contained in the text data can be utilised by an AI that filters out the relevant, deviation-related data from a group discussion (Figure 7.11).

7.5.3 Experience

This approach may sound technically unfeasible and legally impractical in countries with strong privacy guidelines. However, it must be possible to reliably distinguish deviations from other content while also addressing data security concerns.

A few simple measures can dispel the legal concerns:

- Hosting the app on your own servers
- Functional instead of personal accounts
- The user must confirm the saving of texts recognised as relevant.

Modern language AIs are technically capable of reliably differentiating content. For example, in the "CIP" process learning factory at TU Darmstadt, the AI "BERT" developed by Google was trained with labelled chats from production. Thanks to this training, it can recognise when a chat indicates a problem with a component or machine. It then recommends that the chat participants transfer the issue to the dDM system. The users are asked to authorise or refuse the transfer. This also serves to ensure data quality. It has been shown that only relevant and understandable data for other users is transferred in this way [17].

Such a tool for structuring informal processes is particularly helpful when employees work at different times or in different locations and is implemented in five steps [17]:

- Provision of a chat app for employees
- Use of the app for several months to generate a data basis
- Marking of relevant and irrelevant data
- Training the language model for the area-specific language
- Realisation of the interaction skills: Making and saving recommendations in dDM

In combination with a recommendation system (see Section 7.2), the chatbot can provide direct answers to the questions.

7.6 Preparation of Text Data as a Prerequisite for Text Mining

"Nice idea, but it can't work in my production area. The quality of the text data is far too poor." This is a common reaction to the recommendation to use text mining methods in production. However, spelling mistakes, abbreviations, and special terms can be recognised automatically and eliminated where necessary.

7.6.1 Initial Situation

In every production area, there is a specific "slang" consisting of product and equipment designations and their abbreviations. In addition, little attention is often paid to the grammatical completeness of sentences or spelling during the deviation statement. Due to different spellings, abbreviations, and synonyms, it can happen that two deviation statements that mean the same thing do not overlap in one word:

> "ZP 9 not O.K. Screen flickers at system startup. HDMI socket defective."

and

"ZP9 not OK Monitor flickres after start. HDMI-socket def."

These characteristics of text data in production make it difficult to analyse. They must be considered when selecting the analysis methods, but simple measures can reduce the negative effects [18].

7.6.2 Solution Approach

AI models are usually trained with Wikipedia or large collections of books, so they are more familiar with detailed texts. They are also unable to recognise domain-specific abbreviations. If you want to recognise the content of a description with the help of AI, it is necessary to "show" the model the new expressions. This is time-consuming but makes interpreting semantics possible (see Section 7.5). For most use cases in production (see Sections 7.2–7.4), it is only necessary to determine the similarity between the descriptions. However, this is susceptible to spelling errors and different abbreviations.

Analyses and tests have shown hardly any spelling errors in the important words of the deviation statements. Although automatic spelling correction can improve the text data somewhat, this is not important for the results of the analyses. It is more important to recognise synonyms by standardising abbreviations and special characters. The above example sentences become the following:

"ZP9 not OK. Screen flickers on system startup. HDMI socket defective."

resp.

"ZP9 not OK. Monitor flickres after start. HDMI socket defective."

Even without correcting the spelling mistake or replacing the synonyms screen/monitor or start/system startup, five words overlap, which means that the similarity of the two descriptions can be sufficiently determined.

7.6.3 Experience

Tests have shown that correcting all errors and synonyms in the deviation statements is unnecessary when implementing text mining methods with very good results. Abbreviations and special characters can be easily found and corrected in four steps to increase the text data quality sufficiently:

- Search for all terms under five characters or with special characters in all descriptions.
- List these terms by frequency.
- Manual screening of the 100 most common terms for synonyms.
- Replace synonyms in all documents.

The relevant text data can be sufficiently analysed with software support in around 30 minutes. As language usage changes regularly, the procedure should be repeated annually to maintain good results in the text analysis [19].

References

[1] Longard, L.; Meißner, A.; Müller, M. et al.: Digitales daily management – Wohin geht die Reise? *ZWF Journal for Economic Factory Operation* 115 (2020) 9, pp. 645–649.

[2] Eaidgah, Y.; Maki, A. A.; Kurczewski, K. et al.: Visual management, performance management and continuous improvement. *International Journal of Lean Six Sigma* 7 (2016) 2, pp. 187–210.

[3] Burdensky, D. M. A.; Kneissl, B.; Alt, R.: *Descriptive analysis of key figure relationships*. Leipzig Munich: Multikonferenz Wirtschaftsinformatik 2018.

[4] Cohen, J.: *Statistical power analysis for the behavioural sciences*. Hillsdale, NJ: Lawrence Erlbaum Associates 1988.

[5] Zhao, W. X.; Zhou, K.; Li, J.; Tang, T.; Wang, X.; Hou, Y.; Min, Y.; Zhang, B.; Zhang, J.; Dong, Z.; Du Yifan, Yang, C.; Chen, Y.; Chen, Z.; Jiang, J.; Ren, R.; Li, Y.; Tang, X.; Liu, Z.; Liu, P.; Nie, J.-Y.; Wen, J.-R.: A survey of large language models. 2023. https://arxiv.org/abs/2303.18223

[6] Wang, Y.; Chytraeus, J.; Metternich, J.: *Large language models for deviation management*. Wiesbaden: WGP Jahreskongress, Springer 2024.

[7] Lewis, P.; Perez, E.; Piktus, A.; Petroni, F.; Karpukhin, V.; Goyal, N.; Küttler, H.; Lewis, M.; Yih, W.; Rocktäschel, T.; Riedel, S.; Kiela, D.: Retrieval-augmented generation for knowledge-intensive NLP tasks. *Advances in Neural Information Processing Systems* 33 (2020), pp. 9459–9474.

[8] Barrasa, J.; Hodler, A.; Webber, J.: *Knowledge graphs*. Sebastopol: O'Reilly Media, Incorporated 2021.

[9] He, L.; Jiang, P.: Manufacturing knowledge graph: A connectivism to answer production problems query with knowledge reuse. *IEEE Access* 7 (2019), pp. 101231–101244.

[10] Wang, Y.; Bretones Cassoli, B.; Metternich, J.: *Building a knowledge graph from deviation documentation for problem-solving on the shop floor*. Hannover: Publish-Ing 2023.

[11] Nasar, Z.; Jaffry, S. W.; Malik, M. K.; Named entity recognition and relation extraction. *ACM Computing Surveys* 54 (2022) 1, pp. 1–39.

[12] Künzel, H. (ed.): *Erfolgsfaktor Lean Management 2.0*. Berlin, Heidelberg: Springer 2016.

[13] Zaidat, A.; Boucher, X.; Vincent, L.: A framework for organisation network engineering and integration. *Robotics and Computer-Integrated Manufacturing* 21 (2005) 3, pp. 259–271.

[14] Longard, L.; Brungs, F.; Hertle, C. et al: Reduced rework through data analytics and machine learning – A three level development approach. *SSRN Electronic Journal* (2021). 11th Conference on Learning Factories, CLF2021.

[15] Wang, Y.; Longard, L.; Hertle, C. et al.: *Beyond Pareto analysis: A decision support model for the prioritisation of deviations with natural language processing*. Hanover: Publish-Ing 2023.

[16] Müller, M.; Frick, N.; Metternich, J.: Wissen aus betrieblichen Chats nachhaltig nutzen/Results of the transfer project "text analyses in company practice – TexPrax". Sustainable use of knowledge from company chats. *Workshop Technology Online* 111 (2021) 01–02, pp. 93–96.

[17] Müller, M.; Lee, J.-U.; Frick, N. et al.: Extracting problem related entities from production chats to enhance the data base for assistance functions on the shop floor. *Procedia CIRP* 103 (2021), pp. 231–236.

[18] Müller, M.; Metternich, J.: Production specific language characteristics to improve NLP applications on the shop floor. *Procedia CIRP* 104 (2021), pp. 1890–1895.

[19] Müller, M.; Longard, L.; Metternich, J.: Comparison of preprocessing approaches for text data in digital daily management systems. *Procedia CIRP* 107 (2022), pp. 179–184.

Index